The Destr
of Sodom:
What We Have Learned from Tall el-Hammam and Its Neighbors

M000189365

Phillip J. Silvia, Ph.D.

Trinity Southwest University Press
Albuquerque, New Mexico

TRINITY SOUTHWEST UNIVERSITY PRESS
Albuquerque, New Mexico

The Destruction of Sodom: What We Have Learned
from Tall el-Hammam and Its Neighbors
by Phillip J. Silvia, Ph.D.
Copyright © 2016 Phillip J. Silvia

ISBN-13: 978-0692609699
ISBN-10: 0692609695

Cover photo my Michael Luddeni.

ACKNOWLEDGEMENTS

It is with a profound sense of gratitude that I express my appreciation to the many individuals who contributed to the completion of my dissertation research project, which is the foundation of this book.

First and foremost, to my wife of over 45 years, Yvonne, who first inspired and encouraged me to embark upon that journey. Her patience with me as I verbally described my project to anyone who asked would make the patience of Job pale in comparison. She has heard the main themes many times, constantly encouraging me to be more concise. Hopefully, her advice has carried through to the writing of this book as it did with my dissertation. It is to her that this book is dedicated.

To my colleague and doctoral dissertation advisor, Dr. Steven Collins, Executive Dean of Trinity Southwest University and Dean of the College of Archaeology & Biblical History, for his constant encouragement, wealth of knowledge on the Ancient Near East, and decades of experience in Levantine archaeology and ceramics typology. I sincerely appreciate both the information that he shared and the opportunities that he provided for direct archaeological research by allowing me to serve as a Field Archaeologist and Supervisor for the Tall el-Hammam Excavation Project, of which he is Director. It is impossible to estimate the extent of his contribution to the successful completion of the research project that is behind this book.

To Dr. Stewart Wuest, soil scientist with the USDA, the first person to seriously consider my suggestion of soil depletion as a possible cause of the extended occupational hiatus, for introducing me to his colleague's book *Dirt: The Erosion of Civilization*, which proved to be a turning point in the development of my hypothesis.

To the materials analysis team that conducted a major portion of the examinations that were performed on the many kilograms of sand, soil, ash, and rock samples that I collected for the research project. The individual members of the team are listed in Appendix D, however, special mention must be made of Dr. T. David Burleigh, Professor of Materials & Metallurgical Engineering at New Mexico Tech in Socorro, New Mexico, whose involvement began with the analysis of the vitrified pottery samples brought back from Tall el-Hammam by Dr. Collins in 2006. The analysis team seemingly formed out of thin air in 2014 as word of this project spread from the epicenter of Dr. Burleigh's office through the professional networks and personal acquaintances of the members.

This project could not have been completed without the data that they generated through examination of the samples.

Finally, to the Tall el-Hammam Excavation Project staff and volunteers whose friendship, enthusiasm for archaeology of the Ancient Near East, and collective expertise have been a constant source of encouragement. Their probing questions both at the site and around the dining room tables at the hotel over my first four dig seasons at Tall el-Hammam forced me to reason through details that might have otherwise escaped receiving proper consideration.

To all of you, I offer my sincere thanks for your varied contributions.

PJS

CONTENTS

LIST OF FIGURES

LIST OF TABLES

FOREWORD

When—fifteen years ago—I began intensively exploring and re-searching the Middle Ghor (MG = southern Jordan Valley) with a view to excavating its largest and most important archaeological site, Tall el-Hammam, an interesting phenomenon appeared. Previous work in the area by Mallon, Ibrahim and Yassine, Chang-Ho, Leonard, Prag, Flanagan, and Papadopoulos encountered a similar observation: the LBA was architecturally absent from the stratigraphy of excavated sites. While cities and towns in the MG had flourished continuously from the Chalcolithic Period through the Middle Bronze Age, they were all out of business by the end of MB2 (ca. 1600+/– BCE). Flanagan, excavator of Tall Nimrin, called this occupational hiatus the "Late Bronze gap."

But it was not merely a missing LBA. Settlements actually did not reappear in the eastern Middle Ghor until very late in Iron Age 1, mostly ca. 1000 BCE at the beginning of Iron Age 2. The gap was on the order of six-to-seven centuries long. One might ask, "So what?" But when one realizes that the MG—one of the best-watered locales in the southern Levant—was virtually unaffected by either climate change or socio-cultural phenomena from the beginning of the Chalcolithic Period through the Middle Bronze Age—a period of over 3,000 years!—the LB gap appears odd, if not enigmatic.

When civilization centers across the southern Levant all but disappeared from the landscape ca. 2500 BCE as a result of an extended dry period, the cities and towns of the eastern Middle Ghor not only survived, but thrived, actually increasing in number. Contrariwise, while cities and towns in regions contiguous with the eastern Middle Ghor—to the west in the Cisjordan highlands, to the north in the upper two-thirds of the Jordan Valley, to east in the Transjordan highlands—continued into the Late Bronze Age, the entire civilization of the MG ended abruptly. When everyone else went down (ca. 2500 BCE), the MG sites did not. Toward the end of MB2 (ca. 1600+/– BCE), when practically everyone continued into the LBA, the MG sites met with extinction. There was an explanation, but what was it?

Neither famine nor plague nor warfare has such a targeted effect, and certainly not for six or seven centuries. This was a mystery, to be sure, and one worth solving.

There was, however, another possible connection that I was well aware of, because it was this particular link that had brought me to the

LBA - Late bronze age

southern Jordan Valley in the first place: Sodom of biblical fame (or infamy!). Back in the late 1990s Dr. John Moore and I had begun to discuss the geographical issue of Sodom's location. He then posed a question to me: Don't you think the language of the destruction narrative sounds a lot like an airburst event? At that point in time, I didn't even know what that was. So John loaded me up with articles from scientific journals, and I began to educate myself about the phenomenon. Later I called John to say: You're right!

Eventually I was convinced that the Genesis 19 description of Sodom's destruction was about as good a phenomenological description of a cosmic airburst as one could imagine.

When our excavation started at Tall el-Hammam—the site I had identified as Sodom based on geographical details embedded in the biblical text—in late 2005 we were thus very sensitive about observing anything that might have to do with the MB2 destruction layer. The glass jars and vials of dark ash began to mount up, as did the pieces of pottery with melted outer surfaces. Early laboratory analyses demonstrated that, in each case, the glass was the vitrified surface of the clay body itself (glazes were still two millennia in the future!). And the temperature thresholds necessary to create not only the glass but other associated phenomena were far, far beyond anything that might obtain terrestrially. Volcanic magma was 'cool' by comparison.

A few years into the Tall el-Hammam Excavation Project, Phillip Silvia walked through my office door. Based on mutual interests, the conversation turned to cosmic destruction events. Because of his science background as an electrical engineer, his language describing physical phenomena was certainly more sophisticated than mine. But he also had a Bible background, with a master's degree from Gordon-Conwell Theological Seminary. Thus, the subject of Sodom's annihilation wasn't a foreign concept to him. Within a few weeks, Phil launched into a Ph.D. program in archaeology and biblical history. In the field, he was like a duck in water. And with his interest in cosmic impact phenomena, he was the obvious choice as the point-man for solving the enigmatic, terminal destruction of the ancient Middle Ghor civilization.

Phil also had the rare quality of not guarding his investigation from 'outside' eyes, as so often occurs in the archaeological arena. He welcomed into the research those with relevant expertise, and let them ply their specialties in the process (which remains ongoing). He also spent extended time in Jordan beyond our long field seasons. It has been a long road.

Now, several years later, Dr. Phillip J. Silvia presents, for the first time, a summary of the research that has resulted in what I consider to be a compelling explanation of the correspondence between the physical evidence from the Middle Ghor and the narrative description of Sodom's destruction found in Genesis (not to mention numerous later descriptions of the event found in the Qur'an that provides its 'take' on the biblical language).

This research is far from over. In fact, there is already more material to add to the subject than is found in this book. As I write this, the Tall el-Hammam Excavation Project is entering its eleventh season. Several members of the analysis team from universities and research institutions across the US will be on site getting a firsthand look at this enormous Bronze Age city, and taking additional samples for the purpose of furthering their research into what, at this juncture, appears to be a cosmic airburst, or impact event of some character, that snuffed out civilization in the MG toward the end of MB2.

Dr. Silvia has given us much to think about, and he is to be applauded for the rigor of his research, the clarity of his presentation, and the logic of his conclusions.

Steven Collins, Ph.D.
Dean, College of Archaeology, Trinity Southwest University, Albuquerque, New Mexico, U.S.A.
Director and Chief Archaeologist, Tall el-Hammam Excavation Project, Jordan

AUTHOR'S PREFACE

This book is the result of my research into *The Middle Bronze Age Civilization-Ending Destruction of the Middle Ghor*, the nearly circular plain of the southern Jordan Valley immediately north of the Dead Sea, which was the topic and title of my doctoral dissertation. The academic portions of that dissertation have been replaced with anecdotes of my personal journey that led to my research, and the remainder has been edited to make it more readable for the non-archaeologist. I have also added the correlation of my research to the biblical text into the main body of this rewrite as well as retained the original and fuller version in an appendix.

Archaeological investigation of the Ancient Near East (ANE) has largely ignored the Transjordan (eastern side of the) Middle Ghor (TMG). Relying instead on the survey reports of W. F. Albright (published in 1924 and 1925), Père A. Mallon (published in 1932), and N. Glueck (published in 1951), this region of the Jordan Valley has been associated with the Iron Age and later periods because of the surface finds reported by these three giants of ANE archaeology.

This work incorporates the results of a more exhaustive survey of the TMG that was conducted by K. Yassine in 1975-76 and reports of more recent archaeological efforts conducted since the late 1980s led by K. Prag at Tall Iktanu, J. Flanagan at Tall Nimrin, T. Papadopoulos at Tall Kafrayn, and S. Collins at Tall el-Hammam, all of which collectively present a common theme of established settlement during the Early, Intermediate, and Middle Bronze Ages, simultaneous destruction and depopulation during the Middle Bronze Age, an extended period of non-habitation encompassing the entire Late Bronze Age into Iron Age 1, and an eventual repopulation during the following Iron Age 2.

To this is added my own research and analysis (assisted by active and retired scientists and researchers from New Mexico Institute of Mining and Technology[1], Northern Arizona University, North Carolina State University, DePaul University, Elizabeth City State University, Los Alamos National Labs, Sandia National Labs, and private sector industry) on pottery, soil, sand, ash, and rock samples collected from across the

---

[1] To residents of New Mexico, such as myself, the New Mexico Institute of Mining and Technology is commonly called simply New Mexico Tech (NMT), and I will use this familiar name henceforth.

Transjordan Middle Ghor to provide possible explanations for the destruction event itself, reasons for the extended absence of permanent human settlements, and conditions that eventually allowed for the return of human occupation and permanent settlements. The conclusions drawn herein from the synthesis of this information are my own and are offered with the hope that they will encourage dialog within the archaeological and biblical communities and inspire further research by others.

I make no apologies herein for equating Tall el-Hammam with biblical Sodom. Dr. Steven Collins, director of the Tall el-Hammam Excavation Project (TeHEP), and his co-author Dr. Latayne Scott have sufficiently defended this view in their book *Discovering the City of Sodom* (New York: Howard Books, a division of Simon & Schuster, 2013). As a TeHEP Field Archaeologist and Supervisor with five consecutive seasons of hands-on excavation experience at Tall el-Hammam as of the writing of this book, I concur with their conclusions and thus have no reservations with the Tall el-Hammam = Sodom identity. If you don't share this view, then I ask you to read this book anyway and see if you still disagree by the time you reach the end.

PJS

TERMINOLOGY

The terminology used in Ancient Near East (ANE) archaeology is constantly changing, as is the transliteration of place names. Reflecting these changes, this book uses the word *tall* to refer to an occupational mound instead of *tel* or *tell*, except in direct quotations. Also, for ease of reading, most diacritical marks have been removed from foreign words and place names even when they appear in direct quotations. A complete list of standardized spellings used within this book is provided in Appendix A.

I have used the dates, names, and abbreviations for archaeological periods defined by S. Collins for his excavation of Tall el-Hammam (Collins, Kobs, and Luddeni, 2015). These terms and dates are presented in Appendix C.

Finally, this book uses the scholarly designations of BCE (Before Common Era) and CE (Common Era) instead of BC and AD for date references. Dates may also be expressed in "years before present" (YrBP) or "#-thousand years before present" (#KYrBP).

Chapter 1. THE BACKSTORY

Shortly after graduating from Northeastern University in Boston, MA, in 1972 with a Bachelor of Science in Electrical Engineering, I was introduced by an astrophysicist co-worker to the writings of Donald Patten. I subsequently became fascinated by Patten's descriptions of catastrophic themes in the Bible, and that began a 40-year-long hobby of reading a wide variety of books and articles on cosmic catastrophism.

In 1981, I began preparing for a second, parallel career as a pastor and theologian by enrolling in a Master of Divinity program at Gordon-Conwell Theological Seminary in South Hamilton, MA. While studying the Psalms in one of my Old Testament classes, I asked my professor what he thought David meant when he wrote in Psalm 97:1-5 (italics emphasis mine):

> [1]The LORD reigns, let the earth be glad;
>   let the distant shores rejoice.
> [2]*Clouds and thick darkness* surround him;
>   righteousness and justice are the foundation of his throne.
> [3]*Fire* goes before him and consumes his foes on every side.
> [4]His *lightning* lights up the world;
>   the earth sees and *trembles.*
> [5]The *mountains melt like wax* before the LORD,
>   before the LORD of all the earth (NIV).

My seminary professor stated rather matter-of-factly, "That's just poetic hyperbole. This is a Psalm, after all!" But then I found almost the identical expression in Micah 1:3-4, the writings of a prophet that is not in poetic form (emphasis again mine):

> [3]Look! The LORD is coming from his dwelling place; he comes down and treads the high places of the earth. [4]The *mountains melt* beneath him and the *valleys split apart, like wax before the fire, like water rushing down a slope* (NIV).

It became clear to me that phrases like "earth trembling" and "valleys splitting apart" were descriptive of severe earthquake activity. At that time, I thought that "mountains melting like wax" was descriptive of volcanic activity and lava flow. My first season as a volunteer excavator at Tall el-Hammam, however, caused me to change that opinion when I realized that there are no volcanic cones or fissures in the Middle Ghor from which lava might have flowed during the time of either David or

1

Micah. That's when I realized that "mountains melting like wax" is directly tied to the "earth trembling" and "valleys splitting apart." I thus concluded that both David and Micah were probably describing landslides triggered by earthquake activity. From a distance, a landslide makes the hillside look as if it is melting! Furthermore, I became convinced that they were describing what they actually witnessed, using their available vocabulary within the limits of their scientific understanding of the phenomena they were observing.

King David apparently experienced great and severe geophysical and astrophysical phenomena. 2 Samuel 22 contains a song that is attributed to him in which he declares "the Earth trembled and quaked" and "the foundations of the heavens shook" (verse 8). Earth "quakes" we understand, but to say that the Earth "trembled" seems to mean something else, perhaps a wobbling or precession of the Earth on its rotational axis. If this is so, then astronomical references (the locations of stars) would also appear to shift, thus giving rise to the expression, "the foundations of the heavens shook." David also describes great bolts of lightning (verses 13 & 15) and accompanying loud thunder (verse 14) which, in the context of his song, seem to be far beyond what we would expect from the thunderstorms we experience today. He also describes (verse 16) the "valleys of the sea" being exposed and "the foundations of the Earth" (i.e., the land) being laid bare (this is exactly what happens as a tsunami approaches the shoreline) and later (verse 17) being drawn out of the deep waters.

At any rate, such musings over these and similar phrases in the Bible continued to rattle around in my head and would occasionally pop out in conversation with my wife and friends. Many years later, as I approached my 61st birthday, my wife suggested that I finally do something with all of my independent study on catastrophic language in the Bible and "go work on a doctorate or something." Thus, I set about looking for a school where I could "retool for retirement" by acquiring a Th.D. (Doctor of Theology) with a proposed dissertation topic of "Catastrophic Language in the Old Testament and How Catastophism Helped Shape the Theological Development of Ancient Israel."

The Dean of Students at one of the schools I contacted suggested that I pursue a Ph.D. in archaeology instead, explaining that I would be more likely to find the necessary data to support my hypothesis of Old Testament catastrophic language in archaeology than in the biblical text alone.

2

Feeling somewhat confused by this suggestion, I nonetheless began looking for a school where I could do this, and I ended up in the office of Dr. Steven Collins at Trinity Southwest University in Albuquerque, New Mexico, for what I thought would be a half-hour pre-admission interview.

For the first hour-and-a-half he plied me with questions about my interests and previous study in my proposed subject area. For the next half-hour he told me about the Tall el-Hammam Excavation Project (TeHEP) and his conviction that Tall el-Hammam is biblical Sodom. When he finished, he affirmed the recommendation that I should pursue a Ph.D. in archaeology and suggested that I save the second half of my proposed dissertation topic for later, concentrate on the first half, and use Tall el-Hammam for my test case. Thus began my unintentional segue into a "retirement career" as an archaeologist.

Upon my first visit to Tall el-Hammam in Jordan for TeHEP Season Seven (2012), I was immediately "hooked" on archaeology. Seeing first-hand the ancient walls and artifacts convinced me that this was where I was supposed to be looking for answers. The problem was, however, that I was not sure at that time what questions I should actually be asking.

After my second season at Tall el-Hammam in 2013 (Season Eight), I realized that my topic of "Catastrophic Language in the Old Testament" was too broad. I suggested to Dr. Collins that I should narrow my focus to the destruction of Tall el-Hammam. We finally agreed that I should also include the sites in the near vicinity of Tall el-Hammam, and my research into *The Middle Bronze Age Civilization-Ending Destruction of the Middle Ghor* was launched.

As Season Nine (2014) approached, I began making plans to stay in Jordan for an additional six weeks following the six-week dig season to do research at the American Center for Oriental Research (ACOR) and the Department of Antiquities (DoA) in Amman. Not wanting to stay home alone for three months, my wife Yvonne decided to join me for the dig season and extended stay. Little did she know that she had just volunteered to be my de facto research assistant!

Just two weeks before we left for Season Nine, Robb Hermes, a retired scientist from Los Alamos National Labs in New Mexico, visited with Dr. T. David Burleigh at New Mexico Tech (NMT). Dr. Burleigh had been assisting TeHEP with some of the early examinations of materials from Tall el-Hammam collected during previous seasons, and I had left three vials of ash from Tall el-Hammam with him for preliminary examination under the powerful electron microscope at NMT. Robb noticed the three bottles of ash samples that were sitting on his desk and asked about them. After learning from where they came and that we were investigating a possible cosmic airburst in the Middle East, Robb called me as soon as he arrived home and requested a meeting with me and Dr. Collins, which we arranged for the next day. At that meeting we learned that Robb was connected with a team of researchers who had been investigating cosmic airbursts and impacts around the world with a particular emphasis on the Younger Dryas Boundary (YDB) event from about 12,900 years ago. Never before, however, had they been able to obtain any materials from the Middle East, and they were excited to learn that we had a dated archaeological context, albeit much more recent, "only" 3700 years ago for the estimated time of the Tall el-Hammam destruction event. The bottom line: they wanted to offer help with the materials analysis if we were willing to give them access to the materials. Dr. Collins and I took about three micro-seconds to accept the offer, and thus my association with the YDB team began. A full-blown investigation into *The Middle Bronze Age Civilization-Ending Destruction of the Middle Ghor* was officially underway.

Chapter 2. WHERE TO START?

For far too long there has existed an ongoing debate over the story of Sodom and Gomorrah and the other "cities of the plain" that are named in Chapter 10 of the book of Genesis in the Bible. Part of the debate has centered over the actual existence of these cities. Are they *real* places, or are they not? If they *are* real, then why have they not been found? How would they be recognized if they *are* found?

The lack of credible sites for the locations of Sodom, Gomorrah, Admah or Zeboiim has led many to doubt the veracity of the Genesis account and to reduce the stories to the status of myth and legend. Some might concede that there may be a tangible origin to the tale, but disallow any miraculous element. Whether to prove or disprove the truthfulness of the story, the quest to find the "mythical" cities has persisted for a long time, but no one has been able to find a site for any of them that fits the geographical clues given in the biblical text (indicating the *right place*), matches the period corresponding to Abraham (indicating the *right time*), and represents through its architecture and artifacts a major urban center (containing the *right stuff*).

Many theories have been put forward to identify one site or another as the location of Sodom. Some theories have also suggested a candidate site for Gomorrah. But, every theory that has been proposed by archaeologists over the past 100 years and more has failed to satisfy one or more of the *right place – right time – right stuff* criteria. The candidate sites for Sodom and Gomorrah that are most often proposed are Bab edh-Dhra and Numeira, respectively, but they fail miserably the first two criteria. Both of these sites are near the *south* end of the Dead Sea, and the narrative of Genesis 13 tells us that they must be *east* of Bethel and Ai, which are *northwest* of the Dead Sea. Furthermore, Sodom must be *visible* from Bethel and Ai, and the Judean highlands block any view whatsoever of the southern end of the Dead Sea from Bethel and Ai. Bab edh-Dhra and Numeira are simply *not the right place*. Archaeological evidence indicates that Numeira was abandoned about 250-300 years before Bab edh-Dhra; thus, they were not simultaneously destroyed as the narrative of Genesis 19 requires. Also, Bab edh-Dhra itself was abandoned during the Early Bronze Age, about 300 years before Abraham showed up on the scene. Bab edh-Dhra and Numeira, therefore, are *not the right time*, either.

Based primarily on his study of the biblical narrative, Steven Collins, director of the Tall el-Hammam Excavation Project (TeHEP), concluded that Sodom and Gomorrah and the other cities of the plain had to be directly *east* of Bethel and Ai, which necessarily placed them *north* of the Dead Sea, and that is where he began his search. In his original monograph *The Search for Sodom and Gomorrah,* Collins detailed the events that led him to first seriously consider Tall el-Hammam to be the most likely candidate site for Sodom. Along with his co-author Latayne Scott, he provided a more detailed and comprehensive presentation of his search in *Discovering the City of Sodom.* I will not recount those details here—you can read his books yourself—but I do concur with his identification of Tall el-Hammam as biblical Sodom. It is the most likely proposal that has ever been put forth, and, for more than a decade, TeHEP has consistently produced evidence that substantiates the Sodom identification. Tall el-Hammam is the *right place* at the *right time* with the *right stuff,* as Collins has been proclaiming for over a decade.

But, *if* Tall el-Hammam is indeed Sodom, then there should be tangible physical evidence to substantiate the inference found in the biblical text that it was the urban center of a prominent city-state consisting of numerous smaller villages and towns with Gomorrah being largest of them. Secondly, there should also be other nearby sites that could be candidate locations for the lesser city-state combination of Admah and Zeboiim and their surrounding villages. Collectively, they would represent a substantial civilization occupying the region during the Middle Bronze Age (the *right time* for the Abraham narratives). Thirdly, there should be evidence at all of these sites that they were simultaneously destroyed, as the narrative of the biblical text requires.

This is where my own personal quest began. Collins set the stage by defining the *right place – right time – right stuff* criteria, and with these criteria to identify Tall el-Hammam as biblical Sodom. My quest was to see if Tall el-Hammam and its neighbors could tell the *right story.* Would there be sufficient evidence to verify the presence of a thriving Middle Bronze Age civilization as well as its sudden termination, and what other things might be learned through this investigation?

The quest to answer these questions began with getting to know Tall el-Hammam by getting personally involved in the excavation of the site during Season Seven (2012) and learning as much as I could about what had been discovered during the previous six seasons since the beginning of TeHEP. Since then, I have returned to Tall el-Hammam every year in late January to participate in the dig season.

I knew that I would have to also learn as much as I could about the many other archaeological sites that are neighbors of Tall el-Hammam in the Transjordan Middle Ghor (TMG). To meet this objective, I made plans to stay an extra six weeks following Season Nine (2014) to do the needed research in the libraries of the America Center for Oriental Research (ACOR) and the Department of Antiquities (DoA) of the Hashemite Kingdom of Jordan in Amman. From there, it would become a matter of following the data and seeing where it would lead.

Chapter 3. THE PROBLEM

That a thriving civilization occupied the Transjordan (eastern) side of the Middle Ghor (MG)[2] during the Middle Bronze Age (MBA)[3] was largely unknown to early archaeologists, historians, and cartographers of the Ancient Near East (ANE). Survey reports by W. F. Albright (1924; 1925), A. Mallon (1932), and N. Glueck (1951) all documented finding pottery sherds and other items dating to the Iron Age lying upon the surface of virtually every piece of high ground, yet none of these giants of archaeology conducted any excavations in the Transjordan Middle Ghor (TMG) to see what, if anything, was to be found below the surface. Hence, the only sites appearing on most maps of the MG are two sites excavated by others: Jericho on the western edge of the Cisjordan (western) Middle Ghor (CMG) and dating back to the Pottery Neolithic (PN) period; and Teleilat Ghassul on the southern edge of the Transjordan side, just east of the Dead Sea, and dated to the Chalcolithic Period (CP).

A greater understanding of the settlement profile of the TMG has emerged during the past twenty-five years, primarily through the work led by K. Prag at Tall Iktanu, J. Flanagan at Tall Nimrin, T. Papadopoulos at Tall Kafrayn, and S. Collins at Tall el-Hammam. Excavations at all four of these sites have exposed a thriving Bronze Age[4] civilization that covered—if the distribution of these sites across the area is indeed representative of—the entire region. But the existence of a thriving Bronze Age civilization is not all that their reports reveal.

At Tall Nimrin on the northeastern edge of the TMG, Flanagan (1994) noted a conspicuous absence of Late Bronze Age (LBA) pottery and subsequently coined the term *Late Bronze Gap* to describe it. What

---

[2] *Ghor* is the Arabic term for *valley* and is used exclusively and specifically herein to refer to that portion of the Great Rift Valley from the southern end of the Sea of Tiberius/Galilee to and including the Dead Sea. The *Upper Ghor* is the narrow, first 40 km (approximate) of the ghor immediately south of the Sea of Tiberius. The *Middle Ghor* is the final 25 km of the Jordan River's course where the ghor opens to a reasonably flat and nearly circular plain. The *Lower Ghor* is the Dead Sea basin and includes both the deep northern basin and shallow southern basin which is currently above the level of the Dead Sea and being used as an evaporating pan to harvest minerals from the water of the Dead Sea. The focus of this book is the *Middle Ghor*.

[3] See Appendix C for the names, abbreviations, and date ranges for ANE archaeological ages used herein.

[4] The term "Bronze Age" by itself includes the Early, Intermediate, Middle, and Late Bronze Ages.

is this "gap"? It is a break in the occupation history of the site. Simply put, Tall Nimrin was actively occupied during the MBA, vacant and unsettled during the LBA, and resettled some 600 to 700 years later during the Iron Age (IA).

A similar Late Bronze Gap in the pottery witness at Tall el-Hammam was noted by Collins (Collins, *et al*, 2009b), Director of the Tall el-Hammam Excavation Project (TeHEP): "Surface surveying and excavation reveals occupation beginning at least during the Chalcolithic period (some Neolithic material may also be present) and extending with detectable consistency through the Early Bronze Age, Intermediate Bronze Age and into the Middle Bronze Age (all with associated architecture). Late Bronze Age pottery seems to be systematically absent, and consequently there is no discernible LBA architecture thus far." Although not explicitly stated in the excavation reports from Tall Iktanu (Prag 1990), a cursory review of her reports reveals a similar occurrence of the Late Bronze Gap there as well. Although Papadopoulos (2010) claims to have found limited LBA/IB1 pottery at Tall Kafrayn, he also acknowledges a lack of architectural evidence of LBA occupation at that site.

The only other site in the TMG to receive any significant archaeological attention was Teleilat Ghassul, located about 5 km northeast of the Dead Sea. Unfortunately, virtually nothing is left of that site now because it sits in the middle of heavily cultivated fields. The consensus of archaeologists, however, is that Teleilat Ghassul was abandoned by the Early Bronze Age (EBA) (Hennessy, 1982; Khouri 1988) and, thus, adds nothing to help solve the riddle of the Late Bronze Gap that began in the MBA.

Only these four sites have been excavated and confirmed to have been occupied during the MBA—Tall el-Hammam, Tall Iktanu, Tall Kafrayn, and Tall Nimrin—and all four exhibit a Late Bronze Gap in their occupation history. All of these sites are in relatively close proximity to each other (less than 10 km separates Nimrin in the north from Iktanu in the south, as the bird flies). This similarity between these neighboring sites of an occupational hiatus lasting six to seven centuries begs the question: "Why?" Prior to my undertaking of this study, no systematic effort had been made to offer an answer to this question.

Chapter 4. MY APPROACH

My study examined evidence of the occupational history of the MG and the seemingly pervasive disappearance of the occupying culture at the end of the MBA. The initial working hypothesis was that the occupational termination was sudden, affected the entire MG, and included aspects that prevented resettlement of the area for several centuries. My study, therefore, also included an investigation of possible causes of the destructive end of permanent human settlement in the MG during the MBA that might also explain the observed extended occupational hiatus.

The approach taken for the study was fourfold: (1) conduct a systematic review of the available archaeological and site survey reports on the most prominent site in the TMG (Tall el-Hammam), the three major secondary sites (Tall Kafrayn, Tall Nimrin, and Tall Iktanu), and selected tertiary sites; (2) perform a comparative analysis of the destruction profiles these sites; (3) explore possible scenarios that might explain their simultaneous demise and abandonment; and (4) explore possible scenarios that might explain why these sites remained unoccupied for nearly seven centuries.

The main problem that my research attempted to address was to document the long-misunderstood history of the TMG. After over a century of archaeological surveys and excavations in the TMG, no explanation had been offered for the observed Late Bronze Gap in human occupation. Only the more recent research led by S. Collins at Tall el-Hammam was initiated with an *a priori* assumption that the Late Bronze Gap observed at other sites would also be revealed at that site, and this assumption has been confirmed through ten seasons of annual excavations (2006–2015).[5] My research was inspired by a conviction that the time had come to offer a plausible explanation for the Late Bronze Gap.

My initial research question was: *What happened* to the MBA civilization that occupied the MG? That there *was* a MBA civilization occupying the TMG was largely unknown until the 1980s when Prag began her work at Tall Iktanu. That it *did* come to an end during the second half of the MBA is no longer questioned, thanks to the subsequent work done by Flannigan at Tall Nimrin, Papadopoulos at Tall Kafrayn, and Collins at Tall el-Hammam. All three sites show a clear break in the introduction of new pottery forms during MB2a (ca. 1800-1650 BCE) that lasted for

---

[5] A single LBA building was discovered during Season Ten (2015), which concluded just prior to submittal of this dissertation, and will be discussed in Chapter X.

nearly seven centuries until new pottery forms began to appear during IA2a (ca. 1000-900 BCE). All of these sites show evidence of destruction (virtually no standing MBA architecture and, in some cases, significant layers of ash), but not conquest. *What happened* to bring civilization at all of these sites to an apparent simultaneous termination near the end of the MBA, and why they remained unoccupied for so long, must now be addressed.

My initial research question naturally led to a series of secondary and tertiary questions that also had to be answered in order to provide a complete and convincing picture of the occupational history of the TMG. These secondary questions included:

1. What was the *extent* of the destruction? It was not just a single site that was destroyed, but multiple. Evidence must be gathered from as many of these sites as possible and examined to determine the general footprint of the destruction and the relative intensity of the destructive force(s) across the region. Two related issues will be explored: distribution and regional differences.

1a. Based upon the archaeological evidence, was the *distribution* of the destruction sufficiently widespread to support a claim that the entire MG was affected during the MBA?

1b. What *regional differences* can be observed within the MG, both Cisjordan and Transjordan?

2. What was the *scope* of the destruction? Whereas *extent* focuses on the footprint of the destruction, *scope* addresses the magnitude of the destruction in terms of both specific damages at each site and site-to-site differences.

2a. Based upon the archaeological evidence, what *specific damages* were incurred during the destruction?

2b. What is the *site-to-site comparison* of the damage?

3. What was the *nature* of the destruction event? In other words, what does the combination of *extent* and *scope* of the destruction imply? Was the destruction caused by a succession of isolated events that swept across the MG in rapid succession, or was the destruction caused by a single catastrophic event?

After looking at the MBA destruction data contained in the archaeological reports of sites within the MG, a composite profile of the destruction event needed to be created. Using this profile as a guide, a range of physical phenomena capable of producing such a profile needed to be

11

identified and, from that range, a probable cause suggested that best fits the destruction profile. This, of course, required the answering of more questions:

4. Based upon the available evidence, what *destruction profile* can be drawn to characterize what was found?

   4a. What possible *physical phenomena* might have created the destruction profile?

   4b. Which of the possible phenomena provides the best match to the destruction profile and, thus, could be considered the *probable cause* of the destruction?

5. What were the *consequences of the destruction?* Beyond the physical destruction of the sites, there was a clear termination of human occupation revealed through the excavation and survey reports that lasted for six to seven centuries. In addition to accounting for the physical damage and demise of the then resident human population, the identified probable cause must also account for the extended occupational hiatus in the TMG while also accounting for the eventual resettlement.

   5a. What are the possible *causes of the extended occupational hiatus?*

   5b. What *conditions allowing a return* changed that made people able and willing to return?

Chapter 5. DEFINITION OF TERMS

The original target audience for the results of my research was the archaeological community. My research, however, required me to blend a much wider range of scientific disciplines that introduced many new terms that are not part of typical archaeological vocabulary. The intended audience for this book is much broader, but there may be a lack of familiarity with archaeological terms. The definitions for many of the terms listed below vary widely depending upon the context in which they are used and by whom. These variations notwithstanding, I have provided definitions that reflect the meaning I wish to convey herein. Feel free to skip this chapter if you find word definitions boring.

Airburst. The mid-air explosion of a cosmic (see below) object as it makes its fiery streak through the atmosphere. Fragments of the exploded object may either impact the Earth's surface or burn up in the atmosphere.

Archaeological Ages. Periods of human history (both written and unwritten) that are distinguished by the materials used, the tools and objects created, or the ruling powers. (See Appendix C for the names and dates of the archaeological ages of ANE chronology as used herein.)

Bolide. A large cosmic (see below) object, especially one that has the appearance of a fireball or explodes in the atmosphere (see *airburst*).

Cosmic. Of or relating to the cosmos. *Cosmic* is used herein as a synonymous term for *meteoritic* (see below).

Ghor. The term *ghor* is Arabic for "valley" and is used within the corpus of literature pertaining to archaeology in the ANE to mean any cut in the landscape through which water flows, whether continuously, seasonally, or as the result of a sudden downpour of rain. Often, the term *ghor* is used as a synonym for *wadi*, although this latter term is preferred for cuts in the landscape that are usually dry except during the rainy season. Thus, a ghor may be anything from a large ditch, to a river valley, to the Great Rift Valley itself. Within the context of this book, the term *ghor* will be used only in reference to the Great Rift Valley.

Meteor/Meteoroid. An object, such as an asteroid, that enters and traverses the Earth's atmosphere, but does not impact the Earth's surface. A meteor may burn up in the atmosphere (see *bolide*), super-heat and explode in the atmosphere (see *airburst*), or just traverse through the atmosphere and otherwise pass through it harmlessly, leaving nothing behind but a streak of dust.

Meteorite. A meteor that impacts the Earth's surface.

Meteoritic. Of or relating to meteors or meteorites. May also be used to refer to comet fragments traversing the atmosphere or impacting earth. Equivalent to the term *meteoric* (not used herein).

Meteoritics. The study of meteors, meteorites, bolides, comet fragments, etc.

Tall. The term *tall* is the Arabic word for a mound that is built up by human occupation and is equivalent to the previously used terms *tell* [alternate Arabic spelling] and *tel* [Hebrew], which are still used in Israeli archaeology.

Zor. The term *zor* is used to identify the small, secondary channel in the bottom of the ghor through which the Jordan River meanders in its serpentine course southward between the Sea of Tiberius and the Dead Sea.

Chapter 6. ASSUMPTIONS AND LIMITS

Observations of others as well as my own from excavation reports and personal involvement in the Tall el-Hammam Excavation Project (TeHEP) produced a number of assumptions which led to, as well as guided, my study. In order to complete my study in a reasonable amount of time, however, it was necessary to set limits on what my study would include and boundaries defining what would be excluded.

The assumptions that were in place at the beginning of my study included:

Widespread destruction. The destruction affected not just one isolated location, but has been confirmed in at least four locations that are distributed across the Middle Ghor. The assumption is that every site within the MG that was occupied during the MBA suffered the same catastrophe.

Simultaneous termination. The four confirmed sites of Middle Bronze Age occupation all exhibit the same or similar characteristics suggestive of a common, violent, and simultaneous source of destruction during MB2 that is not associated with conquest. The assumption is that every site within the MG that was occupied during the MBA will also show evidence of terminal destruction during MB2.

Extended occupational hiatus. The "Late Bronze Gap" represents a period of 600–700 years with no evidence of human settlement. The assumption is that every site within the MG that was occupied during the MB2 will also show evidence of an occupational hiatus that extends at least into, if not through, the Late Bronze Age.

Airburst event. There is no impact crater in close proximity to the Middle Ghor. The closest known and confirmed meteoritic impact crater is over 150 km to the southeast—too far for any ejecta to reach the MG—yet melt products that are typically associated with either an impact or airburst event have been found on and near Tall el-Hammam in the MG.

Physical evidence. If a meteoritic airburst is a possible mechanism of destruction, then evidence of it should be found at all of the sites involved.

Directionality. Since the force of an airburst decreases according to the inverse proportion of the distance from the source squared ($1/R^2$), the degree of destruction and damage profile at each site should give clues as to the location of "Ground Zero" below the airburst. The topography of individual sites should also offer clues of directionality.

There were several factors which were de facto limitations on what I was able to include in this study. These limitations included:

Availability and access to survey and excavation reports. Publication of annual and final excavation reports is not always guaranteed. I personally observed signs of possible excavation activity on all of the sites I visited, yet there were no reports of record available for some of these sites from the Department of Antiquities (DoA) of the Hashemite Kingdom of Jordan or published reports in the library of the American Center for Oriental Research (ACOR) in Amman. In these cases, I had to rely upon secondary texts for information about these sites.

Access to Transjordan MG sites. Despite having written permission from the DoA to visit and collect sand and soil samples from any site in the MG, not all of the sites were accessible. Three sites of interest (Teleilat Ghassul, Tall Mweis, and Tall Azeimeh) have been completely obliterated by agricultural and/or quarrying activity. Other sites (Tall Bleibel, Tall Mustah, and Tall Nimrin) were either completely inaccessible or offered only limited access because of military presence.

Subsequent human activity and the ravages of time. Finding new physical evidence of a MBA destruction event proved difficult. The destruction event occurred during late MB2, ca. 1650±50 BCE. Subsequent reoccupation during IA2a (beginning ca. 1000–900 BCE) disturbed and/or buried evidence of that destruction event. At least a millennium of time elapsed between when these sites were finally abandoned and modern occupation was initiated.

There were also purposeful delimitations on what I included in my research. These delimitations were necessary in order to complete my research in a reasonable amount of time, but they were not arbitrary— each defined a boundary for my research. The delimitations included:

Middle Ghor sites. No ubiquitous Late Bronze Gap has been noted at any ANE geographical region outside of the MG. It appears that occupation of sites outside of the MG continued with relatively few interruptions right through the LBA. Therefore, my study was restricted to sites within the MG.

Transjordan. Although Jericho, on the Cisjordan side of the MG, is discussed herein, my study focused on the Transjordan side of the MG.

Middle Bronze Age occupation. The purpose of my study was to investigate the destruction event that occurred during the MBA, specifically during MB2. It was not the purpose of my research not the intent of my dissertation to discuss in depth the full occupation history of all sites mentioned herein.

Not intended to argue the identification of the biblical "cities of the plain." Most readers of my dissertation would be familiar with the biblical references to "Sodom and Gomorrah and the cities of the plain." Although I have personal opinions on this subject, they had no bearing on the study or presentation of evidence, or on the conclusions drawn relating to the subject of my research.[6]

Focus. My study focused on: (1) physical evidence from archaeological reports and material samples collected from the region; (2) analysis of the material samples by myself and my research collaborators; and (3) my interpretation of the evidence and analyses.

---

[6] Whereas that was true for my dissertation, this book is different in that I have integrated references to the biblical narrative into the general flow of discussion.

Chapter 7. FOCUSED LITERATURE SEARCH

The starting point of my research was a focused literature search on topics that were germane to the study. There are three assertions in the title of my dissertation that guided the literature search for the study. The first assertion is in the phrase: "The Middle Bronze Age Civilization"— that a thriving civilization occupied the MG during the MBA, despite the fact that the presence of this civilization in the TMG was unrecognized by archaeologists and historians of the MG until relatively recently.

The second explicit assertion is in the phrase: "Civilization-Ending Destruction"—that something happened to bring that occupying civilization to an end that was not just a minor set-back, but a total annihilation. But, there is also a second part to this assertion, namely, that the consequences of the destruction resulted in a sufficiently long occupational hiatus that the people who eventually reoccupied the TMG ushered in a *new*, or at least *different* civilization.

The third explicit assertion is in the phrase: "Destruction of the Middle Ghor"—that this civilization-ending event involved not just a single site, but an entire region.

These three underlying assertions led to the formation of a hypothesis that is composed of three sub-components, namely, that the destruction of multiple MBA civilization sites was (1) sudden, (2) simultaneous, and (3) suggestive of a single, common event.

The three underlying assertions of the title and resulting working hypothesis guided the literature search in three directions. First, the objects of my study were archaeological sites; therefore, a careful examination of archaeological literature pertaining to the MG was required. Second, the destruction encompasses the entire MG; therefore, a careful examination of geological and meteoritic literature was required to identify potential causes of the destruction having both the means and sufficient magnitude to create the destruction observed through the archaeological data. The rationale for the third direction is more complex, but straightforwardly so.

Three things were required for a civilization to take root in a given location: (1) defensible high ground as a protection against enemies; (2) arable ground on which to grow crops sufficient to feed both the people and their livestock; and (3) reliable water for both the irrigation of crops and consumption by people and animals. This is the "three-legged stool" of occupation. Take away any one of these three legs, and the civilization will either collapse or not take root at all. What was left after the destruction event hit the MG? The high ground still stands and the water sources high on the Transjordan plateau still flow to and through the Ghor. Since civilization remained absent for so long, something must have happened to the ground to prevent its return. The third direction, thus, was a literature search on soil dynamics and regeneration as it affects agricultural use.

Chapter 8. RESEARCH METHODOLOGY

The methodology applied to the research for my study was designed to bring together information from three specific disciplines: archaeology, the science of cosmic impacts, and soil science. The research involved a survey and study of literature pertaining to these three disciplines and microscopic examination of materials collected from Tall el-Hammam and seven other ancient sites within an eight-kilometer radius of Tall el-Hammam. Control samples were also collected from the Waqf as Suwwan meteoritic impact crater about 150 km east of Tall el-Hammam.

The research methods employed for my study included both secondary research through published literature and primary research through active participation in the Tall el-Hammam Excavation Project and analysis of material samples collected from Tall el-Hammam and other sites.

One of the underlying assumptions of my study was that the MBA destruction event depopulated a region and not just a single site. Of all the sites that were occupied during the MBA, active excavation at the time I began my study was being conducted only at Tall el-Hammam. It was therefore necessary to conduct secondary research on the occupation history of other sites through a review of the literature that has been published on these sites in the form of surveys and excavation reports. In some cases, excavation reports were never submitted to the DoA or published in any journals. In most of these cases, however, there exists at least anecdotal reference to the sites from one or more written sources. The bulk of my research on the archaeology of the MG was conducted at the ACOR and the DoA in Amman, Jordan. Additional literature research was conducted at public and university libraries in the United States and via the internet.

Meteoritics is not a part of the curriculum for a degree in Archaeology and Biblical History. Meteoritics and cosmic impacts have been a long-time interest in my personal reading, however, and this knowledge base was supplemented for my study with additional, focused, reading from other written sources beyond my own personal library and via the internet. Private correspondence and conversations with various members of the materials analysis team also provided me with specific and detailed information about the terrestrial effects of meteoritic airbursts and impacts.

In addition to secondary research through published literature, I also conducted primary research through hands-on involvement with the Tall el-Hammam Excavation Project (TeHEP) and personal visits to several archaeological sites in the near vicinity of Tall el-Hammam.

As a Field Supervisor at Tall el-Hammam, my access to that site was unlimited and unrestricted. To confirm the regional hypothesis for the MBA destruction event, I had to acquire permission from the DoA to visit other sites and collect material samples for analysis. Blanket permission was granted to visit every site in the TMG.

I was able to collect ant sand on March 27, 2014, from or near seven of the ten that were on my target list. Of the four that I could not access, one (Teleilat Ghassul) has disappeared from the landscape due to agricultural activity, two (Tall Bleibel and Tall Mustah) could not be accessed because of military presence due to a visit by the King of Jordan, and one (Umm Hadar) has been fenced to prevent direct access. Five (Tall Mweis, Tall Iktanu, Tall Rama, Tall Tahuneh, and Tall Kafrayn) of the seven sites visited had line-of-sight visibility of Tall el-Hammam. Visibility from the other two (Tall Nimrin and Umm Hadar) was blocked by the topography. All seven of the sites visited are within 8 km of Tall el-Hammam, as the bird flies.

Five of the seven sites I visited have changed little since the suspension of excavations. One (Tall Mweis) of the other two has been bulldozed flat to help control flooding during the rainy season, and a significant military post has been constructed atop the other (Tall Nimrin). Islamic graves from about the last 100 years cover the surface of one small, but prominent tall (Rama), but they were already present prior to the limited excavations that were done on its summit. The excavations on Talls Iktanu, Rama, Tahuneh, Kafrayn, Nimrin, and Umm Hadar are still visible, although weather and time have degraded their condition.

The specific material sought and collected from each of these sites was "ant sand," i.e., small piles of sand brought to the surface by ants doing what ants do, namely, dig tunnels deep (up to two meters and more) into the earth. Since I did not have permission to excavate at these sites, the hope in collecting the "ant sand" samples was that these tiny excavators would bring spherules (tiny beads of melted silica and/or metal—evidence of meteoritic airbursts—along with otherwise normal grains of sand. For whatever reason, ant sand was hard to find on Tall Kafrayn, but small burrowing animals provided an ample supply of fresh sand from below the surface.

21

In addition to the ant sand, other materials were also collected for examination by a team of experts in materials analysis. Three different types of materials were collected: ash/soil/sand, pottery sherds, and rocks. Ash samples were collected from only Tall el-Hammam. Some of this ash is residue from fire pits used for cooking, but the bulk of it is from what is suspected to be the MBA destruction layer. The soil is from immediately above and below the MBA destruction ash layer. Most of the sand collected was provided by ants of various sizes, but some was collected from small wadis at the Waqf as Suwwan meteoritic crater. (There were no ants in this remote desert area.)

Although literally thousands of pottery sherds are collected and sorted through each season at Tall el-Hammam, only about 800–1,000 are actually registered and kept as "diagnostic" pieces from rims, bases, handles, or spouts, or containing distinct decorations that are indicators of form and age. During the first nine seasons of excavation (2006–2014), only 52 whole (or nearly whole) vessels had been recovered, and most (60%) of these are from the Iron Age and later.[7] Out of the 9,000+ non-diagnostic sherds that are discarded each year, a total of fewer than two dozen have been found over the first ten seasons that display evidence of extreme thermal damage. These are the pieces that were collected for this study.

I first collected rocks showing evidence of thermal stress from the Waqf as Suwwan meteor crater when I visited it on April 4, 2014, with a group from the Friends of Archaeology and History in Amman. Knowing, after that, what thermally stressed rocks look like, I went back to Tall el-Hammam the following week and collected about a dozen samples from the south side of the Upper Tall. Additional "surface-find" rocks were collected by volunteers during Season Ten.

I personally examined most of the ant sand samples under optical microscopes before distributing them to the analysis team for chemical analysis and examination under digital microscopes and Scanning Electron Microscopes (SEM) at their facilities. I also distributed the rocks and pottery samples to the analysis teams for microscopic examination.

The design of my research project evolved from the starting assumptions from prior excavations in the MG that Collins used to launch the

---

[7] Whole vessel distribution by age: CL (1), EB (8), IB (7), MB (5), LB (0), IA (23), Byz (8). Several additional whole vessels were recovered during Season Ten, but they have not yet been catalogued.

Tall el-Hammam Excavation Project. Those assumptions were expanded by J. Moore to include meteoritic airburst as a possible cause of the destruction event.[8] I used those assumptions, amplified by my own prior study in cosmic catastrophism, to develop a research design that also included a plausible explanation for the extended occupational hiatus that followed the destruction.

The literature search proceeded along three tracks. The first track focused on archaeology of the MG so that I could develop a comprehensive picture of the occupational history of the region and glean evidence relating to the occupational hiatus that spanned the LBA. The second track focused on meteoritics in order to develop a better understanding of the physical effects of meteoritic airburst and impact events for comparison with the physical data collected from the archaeological sites. The third track focused on soil generation for agricultural production to test my theory that soil destruction was a major contributor to the occupational hiatus.

Although I maintained records of the collection sites from which I gathered the ant sand and rock samples, I did not pass on this information to the materials analysis team. The samples that they received were identified only by a number so that their analysis would not be biased in any way. They did know, however, that the bulk samples of ash and soil were collected from Tall el-Hammam. This identification was purposeful because they came from the suspected MBA destruction layer and needed special handling and processing.

I was solely responsible for the correlation of results from the examination of the material samples. This correlation was then reviewed by key members of the materials analysis team for technical accuracy before being incorporated into my dissertation.

No unique instrumentation was required or developed for my study. Some very specialized instruments and processes were used by the analysis team, however, for the examination and analysis of the material samples.

The materials analysis team was composed of active and retired scientists, researchers, and professors from a broad spectrum of specializations (see Appendix D for a listing of the individual members). Their common connection, however, is their interest in meteoritics. More importantly, they are all experienced in researching the residual materials

---

[8]   Personal conversations with S. Collins and J. Moore.

from meteoritic airbursts. Collectively, they have invested countless hours investigating the Younger Dryas Boundary (YDB), which is associated with a suspected cometary airburst event that is dated 12,900 years before present (12.9KYrBP).

Laboratory facilities at several institutions were made available through the members of the materials analysis team for examining the material samples. These institutions included: New Mexico Tech, North Carolina State University, Elizabeth City State University, Northern Arizona University, the University of Oregon, and DePaul University.

The laboratory equipment employed in the examination of material samples included optical microscopes (both with and without digital display and image capturing capability), Scanning Electron Microscopes (SEM) with Energy Dispersive Spectroscopy (EDS), and an Electron Probe Micro Analyzer (Cameca 100). Various methods of particle separation were also employed including floatation, screening, and magnetic particle separation. Chemical analysis was also used to isolate and identify the salts that permeate the soil and bind grains of sand together.

The data used for my study was derived from two sources: written material and physical samples of ash/soil/sand, rocks, and pottery. The primary sources of data from written material were books and articles from both print and online. The specific data sought pertained to the Bronze Age occupational history of the MG, the terrestrial effects of meteoritic airbursts, and processes and timelines of soil generation.

Most of the ash, soil, sand, and rock samples used for this study were collected during and soon after the 2014 excavation season of TeHEP (Season Nine). Additional samples were collected during the 2015 excavation season (Season Ten). Pottery samples of specific interest for this study were collected during the 2014 excavation season. Other pottery and melt rock[9] samples were collected during previous excavation seasons. The pottery samples all came from Tall el-Hammam. The melt rock sample came from the area around Tall Mweis, about 9 km SW of Tall el-Hammam.

---

[9] "Melt-rock" is the term used to describe beads of melted silica sand that are typically found in sandy deserts. They are usually formed by lightning strikes sand and rarely exceed 5 mm in diameter. The melt-rock samples examined for this study were irregular in shape and ranged from about 2 cm to over 10 cm.

Seven bulk ash and soil samples of approximately 1 kg each were collected from the destruction layer in Square LS.42K[10] of Tall el-Hammam. A 2 m deep probe was excavated in this Square in Season Five (2010) which cut through the destruction layer. The samples collected for this study were taken from the north and south sides of the probe. Lesser quantities of ash were also collected in 5 and 25 dram vials from fire pits (cooking sites) in nearby squares. Additional soil and ash samples were collected from the MBA destruction layer in Square UA.7GG during Season Ten (2015).

Ant sand collection sites were noted and recorded by Square on Tall el-Hammam or GPS coordinates at other archaeological sites. The ant sand samples were collected in 5 dram vials. A few ants were collected with each sand sample to assist the material analysts with estimating the depths from which the sand was brought to the surface. Each of the 60+ vials was identified sequentially by number to tie it to the location data. This data was not shared with the material analysts, however, so that their examination of the sand would not be biased.

Of the rocks that I collected from the Waqf as Suwwan meteoritic crater, I selected for examination only those that were similar in appearance to the rocks that I collected from Tall el-Hammam. Each sample was individually bagged and numbered to tie it to the location data, but this data was not shared with the material analysts, however, so that their examination of the rocks would not be biased.

All of the individuals involved in the examination of the materials are known to each other through their common interest and active participation in research pertaining to the YDB studies. Collectively, they have developed processes and procedures for examining soil, sand, and rock samples in the search for proxies (tell-tale markers) of meteoritic airburst and surface impact events.[11] All of the samples were examined and processed according to the procedures and the protocols of their respective institutions.

---

[10] The grid system used at Tall el-Hammam for referencing excavation locations is decoded as follows: the first letter is the Lower (L) or Upper (U) Tall; the second letter represents the Field; the number is the N-S grid line on the west side of the Square; and the final letter(s) is the E-W grid line on the south side of the Square. Thus, each Square is named by the grid line intersection at its SW corner. Due to its massive size, there are over 30,000 6-meter grid squares covering Tall el-Hammam.

[11] Proxies are typically distal ejecta comprised of either target rock or extra-terrestrial impactor material and can be produced by both airbursts and surface impacts.

25

Analysis of the materials occurred at two levels. First, the materials were analyzed within the context of the various laboratories and the interchange between the members of the analysis team from their perspective as collaborators on the YDB investigation. This proved to be a tremendous asset to my study because no new procedures were needed to conduct the investigation of the material samples. Second, the results of the materials analysis had to be interpreted within the context of MBA culture and the evidence that it was annihilated by a single catastrophic natural event that created conditions which prevented peoples from returning to the area for several centuries. It is this latter level of analysis for which I assumed full responsibility.

Chapter 9. WHAT THE RESEARCH DISCLOSED

The Tall el-Hammam Excavation Project (TeHEP), in which I am involved as an archaeologist and Field Supervisor, has produced evidence that points to a destruction event involving extremely high temperatures and concussive force during the MBA. This discovery raised the question of whether or not and how neighboring sites that were occupied at the same time were affected. Would they show a similar destruction profile? The total absence of pottery forms from the LBA and presence of very few pottery forms from IA1 at Tall el-Hammam during the first nine seasons of excavation suggest an occupational hiatus of six to seven centuries following the destruction event. Would neighboring sites also exhibit a similar occupational hiatus?

Answers to these questions were sought through a review of books and reports of archaeological surveys and excavations which provided a broad spectrum of information about the settlement history of the TMG. Part of that settlement history includes evidence a destructive event during the MBA that decimated the occupying civilization and rendered the area unlivable for about three centuries on the western edge of the circular plain that is the MG, and for six to seven centuries on the eastern half of the plain.

Preliminary examination of physical evidence acquired from Tall el-Hammam suggests that the destruction event involved temperature profiles that cannot be achieved through natural terrestrial sources and were beyond the technological capabilities of MBA civilization, including after the start of the Industrial Revolution in the late 19th century. Meteoritic impacts and airbursts are the only known, naturally-occurring sources of the thermal energy required to produce the physical changes observed in the materials recovered from Tall el-Hammam. The closest known meteor crater to Tall el-Hammam is about 150 km to the southeast, too far away to be directly connected to the destruction event. This impact crater is also presently undated. Its maximum age is 100,000 years BP, the estimated age of the alluvial fan through which the impactor penetrated. A group of anthropologists studying the area surrounding the crater have suggested dating it to 8,200 YrBP based on a discontinuity in the diet of the indigenous people.[12]

---

[12] Information regarding the possible dating of the impact event was provided through a private conversation at the American Center for Oriental Research (ACOR) with Prof. Dr. Klaus Bandel from the University of Hamburg, Germany.

The absence of an impact crater dated to the Middle Bronze Age motivated a survey of literature on meteoritic airbursts to provide a framework for determining the potentiality of an airburst as the source of destruction, as such phenomena are now known to be more frequent than crater-forming impacts (Lewis, 1999).

Although the review of archaeological literature revealed that the destruction event encompassed the entire 25 km diameter of the MG, its effects did not reach beyond the rim of the Great Rift Valley to the surrounding plateaus where cultures and urban settlements continued without interruption. What, then, explains the extended occupational hiatus within the MG? Was it just religious superstition? Or, was there a physical reason that prevented civilization from returning to the area after its destruction? These questions led to a survey of literature on soil generation to search for a possible agricultural connection to the extended absence of permanent settlements.

The best that literary research can do, however, is to help craft a better hypothesis about the destruction event. Defending that hypothesis requires physical evidence from the sites presumed to have been affected that either affirms or refutes the hypothesis. To that end, I collected soil, ash, sand, and rock samples from Tall el-Hammam and sand samples from seven other neighboring sites. These materials were subjected to microscopic and chemical analysis by teams of scientists from multiple institutions across the United States who are experienced with cosmic impact forming melt products and have previously analyzed numerous samples collected from a wide variety of geographically distributed sites. This, however, was their first opportunity to examine materials collected from the country of Jordan. I also collected and provided to the analysts sand and rock samples from the Waqf as Suwwan meteoritic impact crater in east-central Jordan to serve as a control reference.

The presentation of data in Chapters 10-14 will begin with a review of selected archaeological sites from the eastern half of the Middle Ghor (Chapter 10). This review will include surveys, excavations, and my visits to Tall el-Hammam and its neighbors. The presentation of the archaeological data will be geographical, starting with Teleilat Ghassul and proceeding from south to north through the TMG. This will be followed by presentations of data on the material samples collected (Chapter 11), the analysis of these samples (Chapter 12), meteoritic airburst phenomenology (Chapter 13) and soil generation and morphology (Chapter 14).

A discussion of the data will be presented in Chapter 15, and an analysis and application of the data will be presented in Chapter 17.

Chapter 10. MIDDLE GHOR SITE OCCUPATION

The Middle Ghor—specifically, the eastern half of this section (the Transjordan side) of the Jordan Valley—has long been recognized as one of the richest and most promising archaeological regions of Jordan. S. Merrill (1883), W. F. Albright (1924, 1925), A. Mallon (1932), and N. Glueck (1951) are but four of the great pioneers of archaeology in the ANE who published reports of their extensive surveys of this region. Their surveys, however, were limited primarily to just surface sherding of the sites, so the general conclusions of their reports were biased by the preponderance of Iron Age surface material[13].

Merrill's 1875-1877 survey of the Jordan Valley covered the regions of Gilead (from the Sea of Galillee/Tiberias southward to the Wadi Nimrin), Moab (specifically, the Plain of Moab, the TMG), and Bashan (the Transjordan Plateau above and east of the TMG). He was in the TMG during the spring of 1886 and described the area as being almost entirely given over to agriculture, with large wheat fields occupying most of the valley floor. The ruins of the TMG interested him the most, however. He noted that very few were located on the flat valley floor. Instead, they were mostly located in the foothills and near springs or active flows of water within the numerous wadis. Most of these ruins were located just off the main routes leading north and south with ample water issuing from the hills behind them and abundant fertile fields spreading across the plain before them. He also noted, however, that the entire flow of water from the large stream in the Wadi Nimrin had been diverted for irrigating wheat fields, thus leaving the wadi completely dry.

Glueck (1951, p. 366) provided a good description of the MG as it appeared in the late 1940s:

> The 'Arbôth Mô'áb, the Plains of Moab [i.e., the TMG], have something of the shape of a truncated harp , the N side being formed by the Wâdī Nimrîn, the E by the curving line of hills, and the S by the Wâdī el-'Aẓeimeh close to the NE end of the

---

[13] Sherding is the process of collecting the surface pottery at an archaeological site in order to determine an approximate chronological profile of its occupation history. Unfortunately, conclusions drawn from surface sherding are often inaccurate. Sherding may be a good indicator of what is found on the surface, but it is not a good indicator of what may lie beneath the surface.

Dead Sea. This area is watered [when this was originally written] by a number of perennial streams emerging from the eastern hills, which flow in fairly shallow beds westward across the plains to the Jordan. [Virtually all of these streams have now been dammed for irrigation and flood control.] The three main streams, from N to S, are the Wâdī Nimrîn, known as the Wâdī Sha`ib until it emerges from the hills, the Wâdī el-Kefrein, and the Wâdī er-Râmeh, known in the hills as the Wâdī Ḥesbân. Two-thirds of the way across the plain, the Wâdī el-Kefrein joins the Wâdī er-Râmeh. Further S is the dry Wâdī Ejrefeh, which carries water only in the winter and spring spates, and then to the S of it the dry Wâdī eṭ-Ṭerfeh. To the S of it is the Wâdī el-`Aẓeimeh, which, before it emerges from the E hills is known as the Wâdī el-Herī. South of the Wâdī el-`Aẓeimeh, the eastern foothills crown in towards the NE end of the Dead Sea, effectively closing off the Plains of Moab. . . . [The] cities of the plain of Moab were fully as early, important, and productive of civilized activities and historical phenomena as any in the central and northern reaches of the Jordan Valley.

K. Yassine led additional teams of surveyors in 1975 (northern Jordan Valley) and 1976 (southern Jordan Valley). The goals of this survey fieldwork were to visit, describe, photograph, locate on maps, and collect artifacts from as many archaeological sites as possible in the east Jordan valley, from the Yarmuk River in the north to the Dead Sea in the south, and from the Jordan River on the west to the first rise of the foothills in the east (Ibrahim, Sauer, Yassine 1976). The reports that were generated from the two successive years of survey work, though comparatively brief, are of greater informational value than what was previously published on this region by the aforementioned trio.

Over 120 archaeological sites within the 250± km^2 eastern half of the MG are listed in the report of the 1976 East Jordan Valley Survey (Yassine 1988). Most of these sites are located near primary and secondary wadis. The shallow Pre-Pottery Neolithic (PPN) sites located near the east foothills were thought by the surveyors to be seasonal encampments. They also noted a very strong Pottery Neolithic (PN) and Chalcolithic presence in the valley. The many Early Bronze Age (EBA) sites were located mainly on the higher hills and represented fortified villages

31

or small cities. Intermediate Bronze Age (IBA) sites were generally located away from the major wadis and were sprawling, unfortified agricultural sites. There was an extensive Middle Bronze Age (MBA) occupation of the valley consisting of large fortified cities surrounded by small or medium-sized villages. The Late Bronze Age (LBA) is not very well represented in the north[14] (except at major cites that were later occupied in the Iron Age) and not represented at all in the south. There are a great many Iron Age (IA) sites in the valley, especially from IA1B, IA1C and IA2, but most seem to have been abandoned by the end of IA2. There are only a few IA3 (Persian) sites, but many sites seem to have been reoccupied in the Hellenistic and Roman periods.[15]

The surveys conducted for my study focused on Tall el-Hammam and its immediate neighbors and relied heavily upon published surveys and excavation reports. I also visited many of these sites and observed the excavation work that had been conducted previously by others.

At 64 acres (25 hectares), Tall el-Hammam is unquestionably the largest site in the MG. Because of its size, strategic location, and the many satellite towns and villages clustered around it, Tall el-Hammam is believed to be the controlling urban center of an extensive city-state complex that encompassed the southern two-thirds of the TMG during the Bronze Age (Collins, *et al*, 2008, 2009a, 2009b, 2010, 2011, 2012, 2013, 2014). Because of its status as the largest of the MG sites, Tall el-Hammam was used as the point of reference and comparison for the presentation of the site survey data.

The approximate locations of the significant[16] sites in the general vicinity of Tall el-Hammam are shown in Figure 1. These sites were selected for my study as a representative sample of the many archaeological sites (nearly ten times the number shown, according to Yassine). Also, the sites not included in my study are comparatively small and have little or no written record of any excavation activity, although they may appear in area surveys. Latitude and Longitude coordinates for all of the sites I have visited (Jericho excepted) are shown in Table 1.

---

[14] "North" in Yassine's report means north of Tall Nimrin and south of Wadi Zerqa. "South" in Yassine's report equates to the Middle Ghor.

[15] There is also evidence of later, limited occupation in the Byzantine and Islamic Periods, but these were not considered since the focus of my study was on the Middle Bronze Age and what led up to it.

[16] In this context, "significant" means "relatively large" or "probable MBA sites."

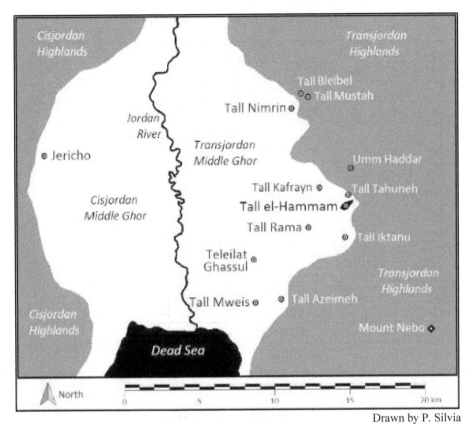

Drawn by P. Silvia

Figure 1 - Significant sites near Tall el-Hammam

33

Table 1 - Site Latitude & Longitude

Site	Lat	Long	City
Teleilat Ghassul	31.805647	35.603042	
Tall Mweis	31.778056	35.604056	
Tall Azeimeh	31.780760	35.623390	Beth-jeshimoth
Tall Iktanu	31.819500	35.671639	Beth-haram
Tall Rama	31.793125	35.604861	
Tall el-Hammam	31.839333	35.672000	Sodom
Tall Tahouneh	31.845889	35.673611	
Umm Hadar	31.862491	35.675707	
Tall Kefrayn	31.850528	35.652944	Gomorrah
Tall Nimrin	31.902389	35.632528	Admah
Tall Mustah	31.906694	35.643583	
Tall Bleibel	31.908972	35.638653	Zeboiim
Tell Jericho	31.871222	35.444167	Jericho
Waqf as Suwwan	31.048808	36.806331	Meteor Crater

The following is a presentation of the data relating to each of these sites.

Teleilat Ghassul

Teleilat Ghassul was located about 7 km SW of Tall el-Hammam in the middle of what is now major agricultural area northeast of the Dead Sea.[17] Despite the fact that it is now virtually impossible to find, Teleilat Ghassul is the most studied and best known archaeological site in the TMG. It is so well studied, in fact, that it became the eponym ("Ghassulian") for the culture of its time period through excavations that were conducted during the 1930s (Mazar 1990, Chapter 3). For many archaeologists working in the Levant, "Ghassulian" and "Chalcolithic" are virtually synonymous terms.

The settlement area of Teleilat Ghassul was over 50 acres (20 hectares), located on a rise that overlooks the NE corner of the Dead Sea. Earliest occupation occurred during the late PN period (before 4600

---

[17] I use the past tense to refer to Teleilat Ghassul because it has been virtually erased from the landscape by present day agricultural activity and could not be located when I attempted to visit the site on 27 March 2014 with DoA surveyor Qutaiba Dasouqi.

BCE). During the Chalcolithic Period (CP, 4600-3600 BCE) Teleilat Ghassul developed from an agricultural village into an apparent administrative center that shows signs of purposeful planning in its layout of long chains of rectangular "broad houses" attached end-to-end. Excavations confirm the original late PN/CP village was not perched on a series of small mounds, but was spread out over a single, flat area (Malon, *et al*, 1934; Hennessy, 1982; Khouri 1988). This lack of occupation mounds at Teleilat Ghassul has undoubtedly contributed to the erasure of the site through present day agricultural activity.

The Ghassulian culture is characterized by its settlement pattern, pottery, stone tools, advanced copper technology, use of art to express religious beliefs, and burial customs. All of these features have no precedent in PN and only a limited continuation in the EBA that followed. Around 3300 BCE, the Ghassulian culture came to an end, and its most important centers throughout the Levant, including Teleilat Ghassul, were abandoned and remained unoccupied in subsequent periods. This sudden cultural collapse marks the transition from the CP to the EBA.

Excavations at Teleilat Ghassul and at other CP sites reveal significant damage from apparent earthquakes rather than conquest. Hennessy (1969) suggested that these widespread earthquakes may have affected water delivered to Teleilat Ghassul through both nearby wadis and underground springs, thereby leading to its relatively abrupt abandonment at the end of the CP.

Conder (1889) described Teleilat Ghassul as "several hillocks of sand strewn with pottery." When Collins and a few others were doing exploration in 2005 before the start of excavations at Tall el-Hammam, they visited Teleilat Ghassul and found a 3.5 cm piece of greenish glass similar to melt rock, but he didn't pay much attention to it. Another team member also picked up some pieces of broken pottery off the surface, which was simply 'desert' surface during MB2, with little or no mound, and would explain its being strewn around the whole area.[18]

---

[18] Personal correspondence with Collins and G. K. Massara, who picked up the random samples.

Tall Mweis

Tall Mweis was located about 2.5 km S of Teleilat Ghassul and about 8.5 km SW of Tall el-Hammam.[19] Tall Mweis was a relatively small town or village on the south side of the Wadi Azeimeh. The occupation mound was on a low rise in a slight bend of the wadi. This site was never officially excavated, but surface sherding prior to the 1980s revealed a dominant presence of pottery from EB1 with some flints (Yassine 1988). This suggests initial occupation during the CP and continuing into EB1. A few sherds from IA2 were also found which suggests a minor and temporary occupation during that period.

Tall Mweis was visited in February 2010 by S. Collins (Director of the Tall el-Hammam Excavation Project). His surface sherding reaffirmed the previous findings. Also found during that visit, however, was a large mass (672 g) of "melt rock" (bonded sand and grit coated with melt glass) that measures about 10 cm in length (see Figure 2), which was extracted from a surface sand matrix next to a stone wall foundation.[20]

---

[19] I use the past tense to refer to Tall Mweis because it was bulldozed flat sometime between January 2009 (when last visited by Dr. Steven Collins and Hussein Aljarrah) and February 2012 (when I last visited the site with Dr. Steven McAllister). I could not locate the site when I attempted to revisit the site on 27 March 2014 with DoA surveyor Qutaiba Dasouqi.

[20] Personal correspondence from Dr. Collins.

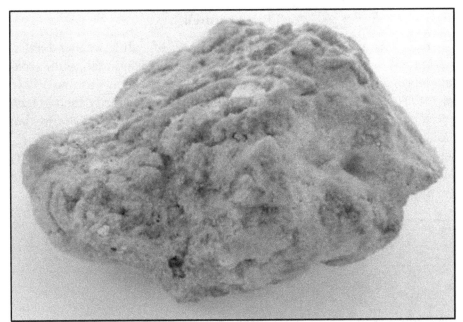

Photo by P. Silvia

Figure 2 - Tall Mweis Melt Rock

I visited Tall Mweis with S. McAllister (TeHEP Field Archaeologist) in February 2012 to see if we could find additional samples of melt rock, but the site had been bulldozed virtually flat since the 2010 visit. We observed several "night digger holes"[21] where the tall had been, but there was nothing to be found in the dirt piles surrounding the holes other than a few unrecognizable body sherds of pottery. An examination of the debris field from the bulldozer surrounding the site yielded only additional sherds of unrecognizable pottery.

From Yassine's data from 1976, it appears that occupation of Tall Mweis began in the late CP and extended through EB1. There is no excavation data available to determine why Tall Mweis was abandoned. Mazar (1990, Chapter 4) speculates that small EB1 sites like Tall Mweis were abandoned as the population concentrated into larger, fortified urban centers during EB2. The abandonment of Tall Mweis may also have been related to the nearly simultaneous abandonment of nearby Tall Azeimeh (see next).

---

[21] As the name implies, "night digger holes" are usually dug under the cover of darkness by people seeking artifacts that they can sell. Their digging causes considerable damage to historical sites and compromises archaeological efforts to study them. Selling antiquities acquired this way is illegal.

Tall Azeimeh

Tall Azeimeh was located almost 3 km E of Tall Mweis and about 7.5 km SW of Tall el-Hammam.[22] It was actually a split tall, with separate North and South occupation mounds. Tall Azeimeh was reasonably large (116 x 75 m) and situated on a high, flat, oval-shaped bench of land overlooking the Wadi Azeimeh which passes between the occupation mounds. Although the site of Tall Azeimeh has been consumed by present-day quarrying activity (see Figure 3), meter-thick foundation walls were still visible when it was visited by Glueck and Yassine in 1949 and 1976, respectively (see Glueck 1951; Yassine 1988).

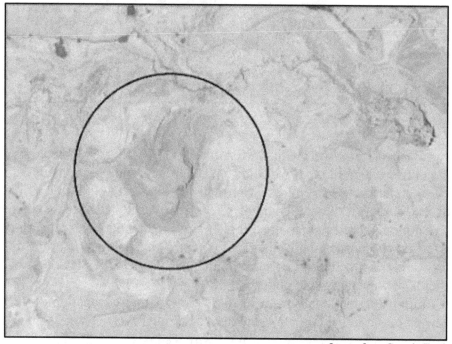

Image from Google Earth

Figure 3 – Limestone quarry where S Tall Azeimeh used to be

Based on the flints found on N Tall Azeimeh by Yassine, it is possible that it was initially settled during the late PN. Both N and S Tall Azeimeh were continuously occupied from CP into EB1. It had a commanding view of the plain to the west, which descends in gradual steps

---

[22] I use the past tense to refer to Tall Azeimeh because it has been totally consumed by a present day quarrying activity due to the limestone deposit on which it once stood.

westward to the NE shore of the Dead Sea. It also controlled a vital water source entering the valley from the eastern hills and an important route linking the valley with Mount Nebo high on the Transjordan Plateau to the east.

Why Tall Azeimeh was abandoned during EB1 is not known. There was no evidence of conquest either on the site or in the vicinity, so it is assumed that something happened (possibly earthquakes) to disrupt its water supply, thereby causing the inhabitants to leave. As was typical of almost every piece of high ground in the Middle Ghor, pottery evidence showed that Tall Azeimeh was reoccupied during IA2, with later sherds extending from the Roman to medieval Arabic periods (Yassine, 1988; Khouri 1988).

Tall Iktanu

Tall Iktanu is about 6 km NE of Tall Azeimeh and about 2 km S of Tall el-Hammam. Like Tall el-Hammam, Tall Iktanu is a double site (see Figures 4 & 5), with a higher tall or acropolis on the NE, and a larger but lower spur to the SW (Prag 1989). Glueck visited the main (north) tall in 1943, but he appears not to have noticed the southern part of the site.[23] Even so, Glueck (1951) considered Tall Iktanu to be the next most important site in the Middle Ghor after Tall el-Hammam. From the top of Tall Iktanu, there is an unobstructed view of the entire MG from the Dead Sea on the SW (see Figure 6) to the spur behind Tall el-Hammam and Tall Kafrayn to the N that blocks the view of Talls Nimrin, Mustah, and Bleibel (see Figures 7 & 8).

---

[23] Conder (1889) also describes only the North Tall.

Photo by K Prag

Figure 4 - Tall Iktanu, looking south, ca. 1990

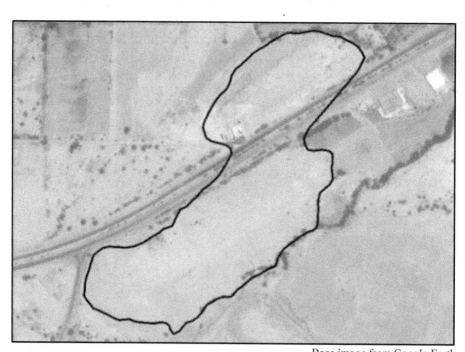

Base image from Google Earth

Figure 5 - Tall Iktanu, split by modern road

40

Photo by P. Silvia

Figure 6 - Looking toward the Dead Sea from Tall Iktanu

Photo by P. Silvia

Figure 7 - Looking toward Tall Rama from Tall Iktanu

42

Figure 8 - Looking toward Tall el-Hammam from Tall Iktanu

Tall Iktanu was the site of a village during the late CP and EB1. After a period of apparent abandonment during EB2 and EB3, Tall Iktanu was extensively reoccupied during the IBA[24] and the settlement area spread out to cover an area nearly 800 m N/S and 600 m E/W. Numerous IBA sherds were recovered by Prag from the North Tall, but she found no architectural remains identifiable to this period. The majority of walls visible on the upper parts of the North Tall appear to be related to an IA fort (Prag 1989).

Rising about 35 m above the plain around it, Tall Iktanu controls the Wadi Hesban as it emerges from the foothills to the east of it and becomes known as the Wadi er-Rameh (Glueck 1951). The North Tall is essentially a natural hill with sandstone outcroppings and jagged ridges

---

[24] Prag (1989) defines the term "EB-MB" as "Intermediate Early Bronze-Middle Bronze" without specifying a corresponding date range. Because she clearly distinguishes the primary occupational phases at Tall Iktanu as belonging to the "EB1, EB-MB, IA, and Persian [IA3]/?Hellenistic periods," I am interpreting her "EB-MB" as meaning "EB4-MB1" (see also Khouri, 1988), which is the equivalent of IB1-IB2, or simply IBA, as defined in Appendix C.

on the N and E sides. The highest ridge supported the IA fort, and the steep sides of the outcroppings and ridges (see Figure 4, above) were incorporated into the defensive walls of the fort (Prag 1989).

Except for a "test pit" and "trench" on the North Tall, Prag confined her excavation of Tall Iktanu to the South (lower) Tall. From her limited excavation of the North Tall, Prag recovered a small sample of EB1 material from the test pit and considerable IA2 and Persian (IA3) material and architecture from the trench (Prag, 1990).

Prag offers no explanation for the abandonment of Tall Iktanu at the end of IB2 and the occupational hiatus that lasted until IA2 in her excavation reports. Khouri (1988) repeats her findings and adds that there is no evidence of violent destruction by fire, earthquake, or attack. Prag (1989, 1990, 1991) concluded that Tall Iktanu was abandoned during both the MBA and LBA. Collins, however, recovered pottery sherds from the surface of the North Tall that were identified as MB2 material when compared with sherds from Tall el-Hammam during the reading by TeHEP pottery experts to whom the origin of the Iktanu sherds was not disclosed. Collins therefore concluded that Prag has missed the MB2 occupation of the North Tall because of her focus on the South Tall.[25] Although there was some level of reoccupation during IA2, IA3, and the Hellenistic Period, Tall Iktanu is one of very few patches of high ground in the TMG where the Romans did not later build an outpost.

Tall Rama

Tall Rama is located about 2.5 km WNW of Tall Iktanu and almost 3 km SW of Tall el-Hammam. Nearly round at the base and rising about 20 m above the surrounding flat plain, Tall Rama in the midst of intensively cultivated, residential area (see Figure 9). From the summit of its conical form, Tall Rama has line-of-sight visibility to the other talls in the southern TMG (see Figures 10 & 11).

In the late 19th Century, traces of foundation walls could still be seen on the sides and summit of the tall (Conder, 1889; Khouri 1988). Those walls have since then been replaced by numerous Islamic graves that can still be seen in various states from relatively new to totally neglected.

---

[25] Private conversations with S. Collins.

Figure 9 - Tall Rama

Figure 10 - Looking toward Tall Azeimeh from Tall Rama

Photo by P. Silvia

Figure 11 - Looking toward Tall el-Hammam from Tall Rama

Both Glueck (1988) and Yassine (1988) did some surface sherding on Tall Rama and reported finding pottery from the IA2, Roman, Byzantine and Medieval Islamic periods, but no authorized excavation of Tall Rama has ever been performed. I visited Tall Rama on 27 March 2014 and observed a survey marker on its summit (see Figure 12) and exposed ashlar walls (see Figure 13) that are consistent with Roman and later occupation, but I was not able to locate any written reports of this excavation work at either the DoA or ACOR. The presence of ashlar blocks is consistent with the Roman sherds previously reported by Glueck and Yassine, but the presence of numerous unworked field stones heaped on the surface (presumably by either grave diggers or grave robbers) suggests underlying architecture from an earlier period. Without reliable pottery evidence, however, it is not possible to date this earlier occupation.

Figure 12 - Tall Rama Survey Marker

Figure 13 - Exposed ashlar block on Tall Rama summit

Since Tall Rama is only 500 meters south of the Wadi er-Rameh (which also runs by Tall el- Hammam), it would have had a reliable source of water during the Bronze Age. Based on this and what has been observed at neighboring sites, I suspect that further excavation of Tall Rama would reveal EB/IB/MB pottery and occupation.[26]

Glueck (1951) considered Tall Rama to be a "fairly high and almost completely natural hill." This, however, is an unlikely conclusion since there are no other geological formations to indicate that the area immediately surrounding Tall Rama is anything other than a flat, alluvial plain. It is more likely that Tall Rama began as an agricultural village, and the entire mound of Tall Rama is the result of continuous human occupation over an extended period of time. Unfortunately, the presence of numerous modern Arabic graves covering Tall Rama will prevent a comprehensive excavation of the site to determine its total history.

---

[26] It is unlikely that further excavation will ever be allowed at Tall Rama because of the many grave sites that now cover its surface.

Tall el-Hammam

Conder (1889) considered Tall el-Hammam to be a natural hillock with an unimportant ruin atop the "mound of the hot bath." Khouri (1988) described it as "one of the largest and most impressive of the unexcavated [sites] in the Jordan Valley." The excavation work by Collins has since revealed Tall el-Hammam to be the largest occupied site in the TMG by far.

Tall el-Hammam is located between two wadis—Wadi Kefrein on the north, and Wadi Ar-Rawda on the south—that flowed all year around prior to their being dammed for irrigation in modern times (see Figure 14). There were also several springs, including two warm (37°C/98°F) mineral springs, surrounding or nearby Tall el-Hammam that provided reliable sources of water. With the fertile plain stretching to the west and south, Tall el-Hammam had everything needed for the development and long-term continuation of a major urban center.

It is clear from Glueck's description of the site that even he did not understand either the true size or significance of Tall el-Hammam. It appears that he noted only the Upper Tall—which represents only about 40% of the total site footprint (see Figures 15 & 16)—and relegated the importance of Tall el-Hammam to second place behind the smaller Tall Iktanu. When Gleuck made his hurried visit to the site in 1941, the remains of a strongly-built Iron Age fort were still visible along with an enclosing fortification wall and the remnants of massive towers at both ends of the long and narrow Upper Tall. The walled rectangular area at the summit measured about 140 m by 25 m in the middle, tapering to about 20 m at each end. The fortification walls measured about 1.2 m thick (Glueck, 1951).

Figure 14 - Tall el-Hammam, looking north, ca. 1990

51

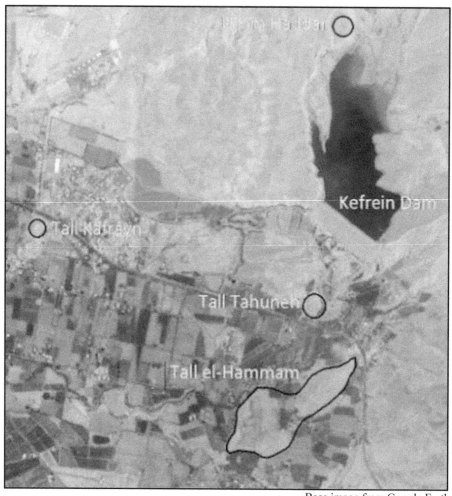

Base image from Google Earth
Figure 15 - Tall el-Hamam and Neighbors

52

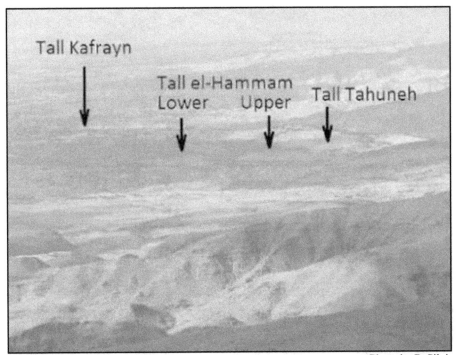

Figure 16 - Telephoto View from Mount Nebo

Although largely ignored by archaeologists, Tall el-Hammam's size, strategic location on the eastern edge of the TMG, and commanding views of the entire Middle Ghor were noted and exploited by various armies during the 20th Century (see Figures 17 & 18). An Ottoman garrison occupied Tall el-Hammam during World War I (1914-18) (Mallon 1932). Considerable damage was done to the site in 1967 when a trench and at least seven defensive gun or tank emplacements were bulldozed across the full length of the Upper Tall. A military minefield covering the saddle between the Upper and Lower Talls and the eastern half of the Lower Tall prevented Prag (1991) from fully examining the site in 1990.

Photo by P. Silvia
Figure 17 - Looking toward Tall Kafrayn from Lower TeH

Photo by M. Luddeni
Figure 18 - Looking toward Tall Kafrayn from Upper TeH

When Prag visited the site in 1965-66, the defensive walls surrounding the Lower Tall were still easily discernable. She also noted the cut made through the wall at the SW end of the Lower Tall by the Ottoman Turks in 1916-17 (and documented by Mallon in 1932) to form a roadway into their camp positioned by the spring in the saddle between the Upper and Lower Talls. When Prag returned to Tall el-Hammam in 1990, she observed that the site had become heavily eroded. There was also much evidence of looting by those seeking an army payroll in gold bullion allegedly stashed in a local cave by the Ottoman Turks as they fled the site at the end of WWI (Prag 1991).

With her path to the Upper Tall blocked by a mine field, Prag confined her 1990 excavations and survey to the western half of the Lower Tall. Near the area she chose to excavate was a slit trench attributed by Mallon to the Ottoman Turks. The trench disturbed some of the ancient foundation walls and contaminated the area with assorted modern military gear, apparel, and spent cartridges, but she was still able to recover significant quantities of EB and IB (her EB/MB) sherds, plus surface finds of Roman and Byzantine sherds. She concluded that the settlement of the Lower Tall may have originated in the EB1 period and was probably continuous through EB3, at which time the importance of the site may have been eclipsed by Tall Iktanu during the IB (her EB/MB) period (Prag 1991).

The mine field had been removed by the time Collins did his initial survey of Tall el-Hammam in June of 2002[27], thus enabling him to sherd both the Upper and Lower Talls. Collins returned to Tall el-Hammam in October of 2002 and conducted more extensive sherding of the Upper Tall. He recovered a large quantity of sherds which affirmed the previously reported profiles of EB, IB, MB, and IA occupation. Conspicuously absent from the repertoire of sherds he recovered at that time was evidence of LB occupation (Collins 2002a).

Collins launched the Tall el-Hammam Excavation Project (TeHEP) in 2005. During the first three annual excavation seasons (2005-2007), the TeHEP team focused on the Upper Tall, taking advantage of the military trench to examine exposed architecture (Collins, *et al*, 2008, 2009b). During Season Three (2008), Collins also completed a survey of the EB/IB city wall and discovered that it extended around both Upper

---

[27] Although the minefield was removed by 2002, a military attaché visited the TeHEP team each season through and including Season Seven (2012) to hand out fliers describing and warning about possible mines that may have been missed.

and Lower Talls. He also noted through extensive surface sherding of the Lower Tall that Tall el-Hammam apparently survived the ubiquitous calamity that caused the demise of EBA settlements throughout the rest of the Levant. Although many of those other settlements never recovered, Tall el-Hammam seemingly thrived during the radical climate change of the subsequent IBA, presumably due to its access to multiple reliable sources of water (Collins, *et al*, 2008).

One of the more interesting finds from the Upper Tall during Season Three was a pottery sherd showing evidence of surface melting (see Figure 19). This vitrified sherd—with one surface turned to glass—was found in a sealed MB2 context in the saddle of the Upper Tall and was the first indicator that the conflagration that produced the deep layer of ash associated with this context may have included temperatures that exceed the normal combustion temperatures of organic materials.

Photo by P. Silvia

Figure 19 - Vitrified Pottery Sherd

The focus of excavations shifted from Upper Tall el-Hammam to Lower Tall el-Hammam during Season Four (2009). A 2-meter wide trench running for 30 m along the 28 N/S grid line in Field LA (hereafter referred to as the "LA trench") was excavated to develop a sense of the stratigraphy and phasing of occupation. Sealed loci with a great deal of pottery were found in abundance, and many architectural indicators revealed close relationships between occupational phases. The occupational sequence noted during Season Four began in EB3 and continued uninterrupted through IB and MB1 into MB2. EB3 structures consisted of mudbrick walls on foundations of 2-3 courses of stone. The EB3 structures were partially destroyed by causes unknown and rebuilt with slightly larger IB mudbrick immediately above the EB3 mudbrick. The EB3/IB structures underwent a subsequent destruction resulting in a thick layer (~30 cm) of ash over a mixed matrix of ash and tumbled mudbrick. These structures were rebuilt following the same wall lines during MB1 and used into MB2. All that remains of the MBA at the surface are numerous foundations and associated pottery. All of this led Collins to conclude that, whatever befell the residents of Tall el-Hammam through the EBA, IBA, and MBA, they reorganized quickly to rebuild, refurbish, and re-create their urban environment (Collins, et al, 2009a).

Further excavations and extension of the LA trench by 30 m to the south during Season Five (2010) made it possible to distinguish the EBA city wall from the MBA city wall that had been built over it. This extension of the LA trench to the south also confirmed that the EB2 occupants were the builders of the initial fortification system surrounding Tall el-Hammam and uncovered the full spread of the MB2 fortification system consisting of laid mudbrick with encased, internal field stone stabilizer walls. Bedrock was reached at the southern end of the LA trench along with the remains of two Chalcolithic/EB1 broadhouses. The foundation of one of the broadhouses was cut into the relatively soft bedrock and contained only Chalcolithic sherds. Thus, the Chalcolithic/EB1 phases constituted an unbroken occupation. With no erosional detritus overlaying them, but only horizontal, engineered fill overlaying them to level the area for construction of the EB2 city wall and adjacent roadway, the occupational sequencing of Tall el-Hammam was confirmed to extend from at least the Chalcolithic Period (CP) into MB2 (Collins, et al, 2010).

The width of the LA trench was extended in Season Six (2011) to 12 m over most of its 60 m N/S length. The architecture uncovered during this season, other than the EBA and MBA fortification walls, was identified as being primarily domestic and included a large communal cooking area containing multiple fire pits and hearths. Ceramic reads from sealed and mixed loci revealed no break in occupation from the CP through MB2. The architecture revealed sequences of destruction (probably by earthquake and/or fire) that were immediately followed by reconstruction on the original building footprints along with occasional expansion through the addition of more rooms. Similar sequences of rebuilding were noted in the defensive systems, including major repairs to the EBA fortification wall, possibly during IB1, following a probable earthquake (Collins, et al, 2011).

Expansion of the excavated area of the LA trench from previous seasons during Season Seven (2012) provided a detailed look at the phases of evolution of the fortification system and adjacent domestic architecture. The presence of ash and ash/architectural debris matrix layers measuring up to one-meter deep was seen as evidence of an MB2 terminal destruction of the city (which was also noted during earlier seasons on the Upper Tall) involving a major conflagration. Several penetrations through the fortification walls were also observed, and these led to the discovery of the main city gate and the foundations of its flanking towers and one of (likely) two much larger monumental towers with several courses of mudbrick lying atop its stone foundation (see Figures 20 and 21) (Collins, *et al*, 2012).

Photo by M. Luddeni

Figure 20 - Looking SW across the LA Trench

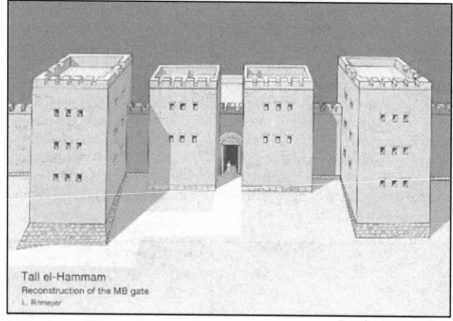

Drawing by L. Ritmeyer

Figure 21 - Reconstruction of the monumental MBA gateway

The MBA gate house wall and pillar foundations were revealed during Season Eight (2013) (Collins, *et al*, 2013), and the IBA and EBA foundations lying below were revealed during Season Nine (2014) (Collins, *et al*, 2014). Excavations during these two seasons at the base of the Upper Tall's north side also revealed numerous storage silos dug deeply into the MBA and EBA rampart walls that were identified with IA2. The need to protect the grain stored in these silos appears to be the impetus behind the construction of the IA2 fortification on Upper Tall el-Hammam.

Season Ten (2015, concluded just prior to submission of this dissertation) provided an unexpected twist with the discovery of destruction remains from a single Late Bronze Age (LBA) building and associated pottery in Square UA.7GG on Upper Tall el-Hammam. During this season, as well as during Seasons One–Three, excavations and soundings within 20 m of this square disclosed Iron Age architecture sitting directly above MBA architecture as has been the case everywhere else at Tall el-Hammam. In this square, the surface IA remains were separated from the LBA remains by more than a half-meter of erosional debris in one corner of the square that tapered off to nothing in the diagonally opposite corner, and the meter-thick LBA remains (center of Figure 22 above out-stretched arm, between and including the dark bands) were separated from the MBA destruction layer (the light band behind the person on the left in Figure 22) by yet another half- meter of erosional debris. Large chunks of charcoal in the LBA destruction layer indicate a slow, oxygen-depleted burn that prevented the timbers from being completely consumed. The dark ash layer within the MBA destruction (the dark band behind the person on the left in Figure 22) is only 3 cm thick and contains only tiny charcoal bits indicating a hotter and complete burn.

Photo by P. Silvia

Figure 22 - LBA and MBA destruction layers

This isolated building does not constitute a settlement. The southern Levant was under Egyptian (18th Dynasty) control during this time. Jericho was resettled during this period, and it is conceivable that an outpost was also built across the Jordan and above the flood plain to monitor caravan traffic in the area during the latter half of the LBA. Found within the destruction debris of this building was a pair of bronze balance scale pans and some weights, which suggests that this building may have been used for collecting tariffs from passing caravans.

A topographical list of outposts and waystations attributed to Tuthmosis III (Aharoni, 1979; Grimal, 1992) contains a list of names that is almost identical to the list of place names identifying the march of the ancient Israelites as they crossed into Moab from the east and worked their way down into the Middle Ghor to their final encampment.[28] Of particular interest in this travelogue are the final few encampments that are listed. From Iye Abarim on the border of Moab they camped at Dibon Gad, Almon Diblathaim, the mountains of Abarim near Nebo, and made their final encampment on the plains of Moab by the Jordan across from Jericho between Beth Jeshimoth[29] and Abel Shittim.[30]

Many scholars and historians of the ANE believe that Lower Tall el-Hammam is Abel Shittim. Tell Azeimeh, 7.5 km SW of Tall el-Hammam, is identified by some (e.g., Aharoni, 1979) as the location of Beth Jeshimoth because of its prominent position on the trade route coming down into the MG from Mount Nebo. With this identification of Beth Jeshimoth, the final Israelite encampment could have been anywhere between Tall Azeimeh and Tall el-Hammam (possible) or spread out between the two (unlikely).

Collins suggests that Beth Jeshimoth may actually be Upper Tall el-Hammam, and that the building found during this most recent excavation season may be the actual site.[31] In this scenario, Moses could have pitched his tent *at* Beth Jeshimoth on Upper Tall el-Hammam—thereby giving him a commanding view of Lower Tall-el Hammam and the surrounding plain—and the Israelites could have made camp on Lower Tall el-Hammam (Abel Shittim) and, if necessary, the flat land in the immediate vicinity of the tall.[32]

Through ten seasons of excavation, the continuous occupation of Tall el-Hammam from the CP through MB2 has been confirmed. The EB2 occupants of the site were the original builders of the extensive fortification systems that surround both Lower and Upper Tall el-Hammam, which were strengthened significantly during EB3. The IBA occupants used most, if not all, of the EBA footprint, including the fortifications.

---

[28] The entire route of the Israelites from Egypt to Abel Shittim is listed in Numbers 33.

[29] The Hebrew name *Beth Jeshimoth* means "house of the wilderness."

[30] The Hebrew name *Abel Shittim* means "mourning place of the acacia trees."

[31] Private conversation with S. Collins.

[32] Whether or not the LBA building was still standing is irrelevant. Egyptian hegemony over the southern Levant had been in decline for almost 40 years (Collins, 2005). The Israelite hoard would have had little opposition to taking over the site.

MBA occupation is strongly attested, particularly in its fortification ramparts and walls on both Upper and Lower Tall el-Hammam, the monumental gateway on the south side of the Lower Tall, and in numerous domestic architectural contexts. Only one structure belonging to the LBA has been found, and none from IA1. The IA2 city is well attested on Upper Tall el-Hammam through both monumental and defensive architecture and to a lesser extent on Lower Tall el-Hammam through primarily domestic and possible cultic architecture. Despite the surprise of discovering the LBA outpost on the Upper Tall, one isolated building does not invalidate the observation of a Late Bronze Gap in permanent settlements. Except for the Roman bath structure near the hot spring in the south side of the saddle between Upper and Lower Tall el-Hammam, no architecture from any later period has been discovered. In addition to a Roman presence, pottery and other material remains indicate limited IA3 (Persian), Hellenistic, Byzantine, and Islamic occupation. Having left no distinct architecture, however, it is assumed that either previous architecture was re-used or the occupants dwelt in temporary structures such as tents.

Tall el-Hammam was one of the largest cities in the southern Levant from the EBA through most of the MBA. It was also the urban hub of a significant city-state that controlled the trade routes coursing through the Middle Ghor from at least EB2 through much of MB2. The scale and strength of its EBA-IBA and MBA defenses attest to a strong centralized government able to maintain its urbascape successfully over a long period of time—at least 1,500 years. Every indication is that it maintained its city-state status throughout the IBA (ca. 2500-1950 BCE), including numerous satellite towns and villages, a phenomenon unique in the southern Levant during that period. Architectural and artistic motifs suggest not only an affinity with certain Canaanite coastal sites (such as Tell Kabri and Ugarit), but also with Minoan Crete (Collins, *et al*, 2013). Following a hiatus with no urban settlement that extended through the entire LBA and IB1, Tall el-Hammam was finally reoccupied during IA2 but largely abandoned by the end of IA3 (the Persian Period).

Tall Tahuneh

Tall Tahuneh (see Figure 23) is only a few hundred meters north of Tall el-Hammam (see Figure 24), just west of the modern Kefrein Dam. At the base of Tall Tahuneh are the remains of an old but undated water-powered grain mill. At the top were the ruins of a Roman fortress that were noted by Mallon and affirmed by Glueck (1951). Although Yassine (1988) attributed the occupation history of Tall Tahuneh to the EBA, MBA/LBA, IA2/3, and later, he also labeled the Bronze Age pottery designations as "possible," thus admitting his lack of precise knowledge of Bronze Age pottery forms at this point. From his own sherding of Tall Tahuneh, Collins (2009b) concluded that a small EBA settlement occupied the site that continued into the MBA. He found no pottery evidence indicating LBA occupation, but concurred with Yassine's IA2 and later occupation assessment.

Base image from Google Earth

Figure 23 - Tall Tahuneh

Figure 24 - Tall el-Hammam from Tall Tahuneh

Umm Haddar

Umm Haddar is located at the north end of the reservoir behind the Kefrein Dam, about 2.5 km north of Tall el-Hammam (see Figure 25). Umm Haddar is not a single, isolated occupation mound, but a number of small hillocks and ridges that line the Wadi Kefrein. The dominant hill is characterized by a rectangular Hellenistic structure (see Figure 26) built of undressed field stones (Waheeb, 1997). Collectively, the distributed occupation sites of Umm Haddar reveal continuous occupation from the Chalcolithic Period through the EBA. There was a later period of occupation during IA1/2. The latest architecture is from the Hellenistic Period, but pottery evidence suggests later reuse and occupation of the Hellenistic structures during the Byzantine Period.

Photo by P. Silvia

Figure 25 - Looking North from Tall el-Hammam

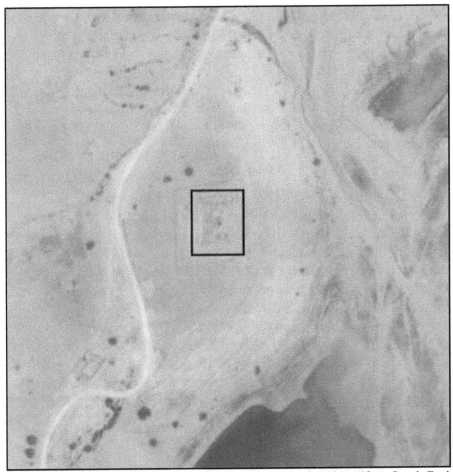
Figure 26 – Hellenistic structure at Umm Haddar

Modern agricultural activities plus the construction of the Kefrein Dam have subjected the site to considerable damage from plowing and bulldozing. According to the survey report of the Wadi Kefrein that was done in 1997 prior to construction of the dam, no settlements were located in the bottom of the wadi or covered by the filling of the reservoir. In other words, all of the settlements of Umm Hadar had been built above the seasonal flood level of the Wadi Kefrein.

M. Waheeb (1997) described the Chalcolithic site as "enormous." Foundations of undressed field stones for numerous walls of square and rectangular rooms were found *in situ*, but in very poor condition. He was unable to determine if these foundations represented a single, massive structure or the attached individual homes of a village setting.

The Wadi Kefrein would have been a reliable water source for the settlements within the immediate vicinity of Umm Haddar during the Bronze and Iron Ages, thus supporting continuous occupation from the Chalcolithic Period through the EBA. Since Waheeb's survey immediately preceded construction of the Kefrein Dam, he had insufficient time to determine the cause of abandonment at the end of the EBA or the reason(s) why an occupational hiatus followed through the Middle and Late Bronze Ages.[33]

Tall Kafrayn

Tall Kafrayn is about 2 km NNW of Tall Tahuneh and about 2 km NW of Tall el-Hammam (see Figure 27). Albright, who did not excavate Tall Kafrayn, described it as being "not a true tell, and probably of not great antiquity" (1925). Glueck, who also did not excavate Tall Kafrayn, more accurately described the site as a "completely isolated, rocky hill, which juts like a tall cone, approximately 35 m. high above the surrounding plain" (1951). According to Khouri (cited without reference), Glueck also reported finding Chalcolithic sherds from the plain southwest of the tall, suggesting to him that a settlement dating to this period once existed nearby. On the tall itself Glueck (1951) found numerous sherds from IA1/2, with none earlier and some from the later Roman and Byzantine periods. Thus, Glueck consigned the initial settlement of Tall Kafrayn to the Iron Age.

---

[33] Waheeb views the Early and Middle Bronze Ages as contiguous and, therefore, does not distinguish the Intermediate Bronze Age where IB1 equals the old EB4 and IB2 equals the old MB1 (see Appendix C).

Base image from Google Earth

Figure 27 - Tall Kafrayn

During the late 19th Century, Conder (1889) reported a "modern Arab tomb" on the summit of Tall Kafrayn. When Glueck surveyed Tall Kafrayn in the late 1940s, he also observed relatively modern Arab graves on the small, flat top of the tall amid the ruins of a small Iron Age fortress along with rock-cut tombs, mostly destroyed, among the ribs of rock at the base of the tall on the west side and similar remnants of tombs on the east side. By the time T. Papadopoulos began excavations on Tall Kafrayn in 2002, the slopes of the tall had been damaged to some extent by erosion and modern military defensive activities (Papadopoulis, 2007). No mention was made in his first (2007) report of Arab graves on the summit of the tall, which suggests that they had been removed by the military subsequent to Glueck's report.

Tall Kafrayn provides a commanding and unobstructed view of the Middle Ghor (see Figure 28). It also controlled the ancient trade route that circuited the eastern edge of the TMG from Tall Iktanu, past Tall el-Hammam and Tall Kafrayn, and continuing northward past Tall Nimrin between Talls Bleibel and Mustah, and up the Wadi Nimrin toward Gilead.

Photo by P. Silvia

Figure 28 - The view from Tall Kafrayn

After several seasons of excavation, Papadopoulos (2010) concluded from architectural and pottery evidence retrieved from the site and pottery evidence at a nearby ancient cemetery that settlement of Tall Kafrayn probably began during the EBA and continued into the MBA. He offers no explanation for the apparent abandonment of the site at the end of the MBA. Papadopoulos concluded that the strategic location of Tall Kafrayn led to the reoccupation of the site and development of an important settlement and fortress near the end of the LBA (based on limited finds of LBA pottery) that continued through most of the IA. Like Prag at Tall Iktanu and Flanagan at Tall Nimrin before him, and Collins working concurrently at Tall el Hammam, Papadopoulos observed that there were no architectural remains to fill the occupational gap between the MBA and IA.

Tall Nimrin

Tall Nimrin is located about 5 km NNW of Tall Kafrayn and about 7 km NNW of Tall el-Hammam. Tall Nimrin cannot be seen from these other two sites, however, because of an interposing ridge. Tall Nimrin is the second largest MBA site in the TMG, about one-fourth the size of Tall el-Hammam. It, too, has some satellite towns and villages and may have also been the controlling urban center of a lesser city-state (see Figure 29). Whether Tall Nimrin and its satellites were an independent city-state or under the hegemony of the larger Tall el-Hammam cannot be clearly determined at this time.

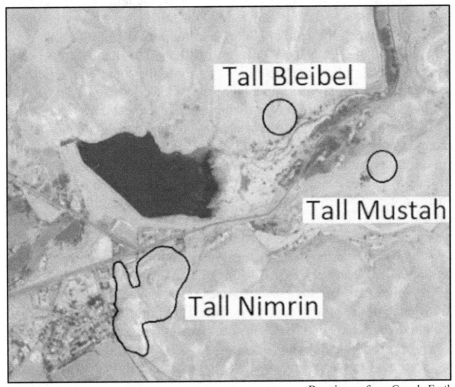

Base image from Google Earth

Figure 29 - Talls Nimrin, Bleibel and Mustah

Although several early archaeologists visited and commented on Tall Nimrin and its surroundings, none excavated the site prior to the work of J. Flanagan in 1989. In the late 19th Century, Conder (1889) observed a substantial flow of good drinkable water from the perennial streams in the wadi on the north side of Tall Nimrin. This flow fed from springs high in the mountains to the north just below the city of Salt. When Glueck surveyed the site in the 1930s, he found large numbers of Roman through mediaeval Arabic sherds, but nothing that he recognized as being from any earlier period (Glueck, 1951). Thus, for a long period of time, the earliest occupation of Tall Nimrin was thought to be from the early Roman period.

In 1982, remnants of a Byzantine church were found on the south side of Tall Nimrin during the construction of a house (Flanagan, et al, 1990, 1992, 1994a, 1994b, 1996). Jordan's Department of Antiquities declared Flanagan's work at Tall Nimrin to be a "salvage" or "rescue operation" because of the encroachment of housing on the lower slopes of the site, military encampments on the top of the tall, and the removal of a large portion of the northern side of the tall to make way for a modern road between the Jordan Valley and the city of Salt via the Wadi Shu'aib (also known as the Wadi Nimrin after it exits the foothills below ancient Gilead). Flanagan was called in prior to the completion of efforts to widen the original road into a highway.

Flanagan's first season of excavation disclosed that the site was originally established on a relatively flat alluvial plain and not upon a natural mound as originally proposed by Glueck. Tall Nimrin, therefore, is composed almost entirely of the accumulated remains of human occupation instead of geological deposits. Flanagan's initial analysis of the 41,000 sherds collected during his first season led him to believe that Tall Nimrin had been continuously occupied from the late EBA to the present. He was not confident of his pottery knowledge at that time, however, and questioned whether he had sufficient justification to support occupation during the LBA and early IA (Flannigan, et al, 1990).

By the end of his second season of excavation, Flanagan's knowledge of the pottery repertoire from Tall Nimrin had increased and led him to conclude that the earliest settlement represented in the lowest layers of the mound dates to IB1[34] through MB2 (Flannigan, et al, 1992). He also identified an approximate 500 year occupational gap spanning the LBA which was followed by reoccupation during the IA. Flanagan noted that the MB and IA2 periods were particularly well-represented at

---

[34] Flanagan calls this period EB4, which is equivalent to IB1 in the chronology defined in Appendix C and followed herein.

the site through at least one meter of IA2 strata and 3.5 m of MB strata that were excavated on the north side of the tall.

During his third season of excavation, Flanagan observed that the IA2 structures were built immediately upon the MB2 structures and re-used many of the MB2 foundations. Although there was much MB2 debris beneath the IA2 occupation, he found no clear evidence of violent destruction to explain the abrupt abandonment of the site at the end of the MBA[35] or the Late Bronze Gap[36] in occupation (Flannigan, et al, 1994).

During his fourth season of excavation, Flanagan determined that the monumental fortification walls surrounding Tall Nimrin were constructed during the MBA occupation and used a base for the later IA occupation (Flannigan, et al, 1996). Between 13 and 15 courses of well-preserved mudbrick were exposed. Additional IA2 construction directly upon underlying MB2 architecture was noted, thus affirming that the Late Bronze Gap in occupation extended into if not through IA1. Also, MB2 construction directly upon underlying EBA architecture affirmed previous findings that the earliest occupation of Tall Nimrin occurred in the late EBA[37].

Tall Mustah

Tall Mustah is located about 1.7 km NE of Tall Nimrin. It is a sharply-defined, flat-topped, wedge-shaped tall that dominates the confluence of the Wadi Nimrin in its north side and the dry Wadi Jari'ah on its south side (see Figure 30). Tall Mustah occupies an exceedingly important strategic position, overlooking the trade route from the TMG to Gilead and the perennial flow of the Wadi Nimrin (Glueck, 1951).

---

[35] Although clear evidence of MBA destruction was found in many areas on Tall el-Hammam, such evidence was clearly absent in Field LS, for which I am the Field Supervisor. In this field, the Iron Age people dug through the MBA destruction layer and removed about two-thirds of the underlying MBA foundation stones, which they recycled in the construction of their own building. Flanagan noted that the IA architecture sat directly upon the MBA architecture, and this could explain the absence of a destruction layer.

[36] Flanagan coined the phrase "Late Bronze Gap" in his Season Three report to describe the observed occupational hiatus between the MBA and IA. Collins has since adopted Flanagan's phrase to describe the identical occupational hiatus observed at Tall el-Hammam.

[37] Flanagan's "late EBA" is EB4 into MB1, which is equivalent to IBA in the chronology defined in Appendix C.

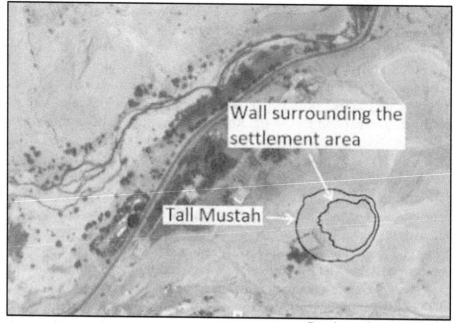

Base image from Google Earth

Figure 30 - Tall Mustah Settlement Area

Among the many EBA pottery sherds found by Glueck during his survey of the TMG in the 1930s, he also found a double-edged flint and a "fan scraper" that are Ghassulian in type. This led Glueck to assign Tall Mustah settlement to the beginning of

the EBA. The east Jordan Valley survey conducted by Yassine in 1975/76 reported finding sherds from the EBA, as did Glueck, along with a few samples from the Byzantine and later periods (Yassine , 1988). He also noted a "few probable LB" sherds, but this identification is dubious because it is not attested elsewhere. During his own survey of Tall Mustah in 2002, Collins reported finding sherds indicating that the site was occupied from at least the EBA into the MBA.[38]

Tall Bleibel

Tall Bleibel is located less than 500 m west of Tall Mustah, across the Wadi Nimrin, standing on the last high terrace of the east hills that face the plain below it to the west (see Figure 31). The strategic location of Tall Bleibel atop this high, isolated hill mirrors the equally strategic position of Tall Mustah across the Wadi Nimrin (Glueck, 1951). No

---

[38] This information was acquired through my personal conversations with Collins.

74

building remains are visible on Tall Bleibel, but the fortification wall surrounding the settlement area is still faintly visible from the air.

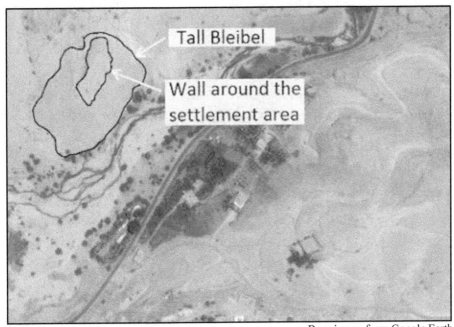

Figure 31 - Tall Bleibel Settlement Area

Glueck reported that all of the sherds found on the surface of Tall Bleibel belonged to the IA. According to his footnotes, however, be never sherded Tall Bleibel himself, but relied on reports from Albright, Abel, *et al*, for his assessment. During his own survey of Tall Bleibel in 2002, Collins reported finding sherds indicating that this site, like Tall Mustah, was occupied from at least the EBA into the MBA.[39]

Other Sites

K. Prag and H. Barnes (Prag & Barnes, 1996) investigated three fortresses on the Wadi Kefrein near Tall el-Hammam in 1995 (see Figure 32). One of these, Tall Barakat, is a little more than 1 km north of Tall el-Hammam and a little over 0.5 km north of Tall Tahuneh, and shows evidence of EBA followed by Roman occupation with nothing in between.

---

[39] This information was acquired through my personal conversations with Collins.

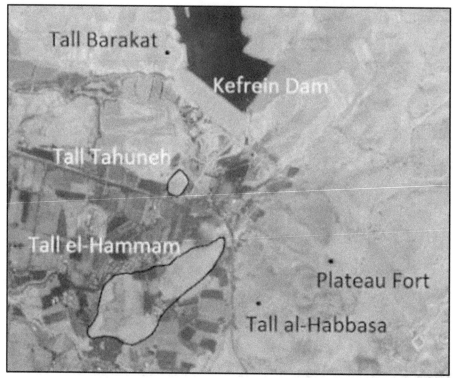

Base image from Google Earth

Figure 32 - Three Fortresses near Tall el-Hammam

Another, Tall al-Habassa, is very close to a hot spring at the east end of Tall el-Hammam and is attributed to the Roman Period because of the aqueduct that channeled water toward the Roman bath house on the south side of Tall el-Hammam. Of interest to this study is the "Plateau Fort." This site is less than 1 km east of Upper Tall el-Hammam and was previously visited by Glueck (1951) and described in significant detail by Mallon (1933). The earliest sherds noted by Mallon were from the EBA, and he concluded that the fortress dated to that period and should be associated with the builders of the many dolmens that dot the surrounding hillsides. The surface sherding conducted by Glueck led him to conclude that the fortress was the product of IA builders. Based on the relative proportions of pottery evidence that she recovered, Prag agreed with Mallon that the site was in use during the EBA, but she concluded that Glueck was probably correct in attributing the fortress to IA builders. What is significant about the Plateau Fort is that pottery sherds were found from the EBA, MBA, IA, Roman, and Byzantine periods, but not

76

from the LBA, which is consistent with the pottery profile seen at Tall el-Hammam and its other neighbors.[40]

---

[40] Prag does not identify EB4 and MB1 as more the recently defined IB1 and IB2, respectively; therefore, her reference to EBA and MBA includes the IBA.

Chapter 11. MATERIAL SAMPLES

No standing structures from the MBA remain at any of the sites in the TMG. Only stone foundations and limited sections of mudbrick walls bear witness to the architecture that once served as housing and public buildings. With few exceptions, mudbricks exist only in the remains of massive ramparts surrounding Tall el-Hammam and the monumental gate towers. Limited fallen patches of mudbrick superstructure walls have been found, but most are broken or crushed. The vast majority of recoverable material is pottery, and most of that is in the form of broken pieces. These pieces are, in fact, quite important, because they provide clues through their form, material, and firing/hardness that enable archaeologists to assess the date of the context in which they are found.

The architectural remains (stone foundations and limited sections of mudbrick walls) provide clues regarding the magnitude and scope of the destruction, but not necessarily the mechanism of destruction. For example, walls can fall because of earthquake as well as blast, and fires can result from either. The meter-thick layer of ash in some locations on Tall el-Hammam suggests a major conflagration, more than one would expect from open cooking fires or oil lamps overturned by earthquake. To tell the difference, microscopic analysis of the materials is required with instruments far more sensitive than a typical optical microscope. Therefore, I set about collecting materials from Tall el-Hammam and other sites during an extended stay in Jordan from January 25 to April 18, 2014, and brought those samples back to the U.S. for laboratory examination and analysis.[41]

Soil and Ash Samples

Prior to departing for Jordan, I had originally planned to bring back mostly relevant pottery and ash/soil samples. Just two weeks prior to leaving, however, I had the good fortune to be introduced to R. Hermes, a retired scientist from Los Alamos National Labs and recognized expert in trinitite, the radioactive glass formed by the explosion of the first atomic bomb at the Trinity Site in New Mexico. He suggested that I take advantage of the excellent excavation skills of ants and collect their sand piles from wherever I see them, the hope being that they would bring to

---

[41] I must express my thanks to A. West for orchestrating the shipment of samples back to the United States through his associates at the University of California at Santa Barbara under their USDA permit.

the surface microscopic beads of glass formed from melted silica or other minerals. These melted beads of glass, or "spherules," are one of several signature markers of meteoritic airburst and impact events (Bunch, *et al*, 2012). If such spherules exist in the MBA destruction layer at Tall el-Hammam or elsewhere, then Hermes believed that ants would bring them to the surface along with regular grains of sand.

As a Field Supervisor at Tall el-Hammam, I had unlimited access to the site and was able to collect 33 5-dram vials of ant sand from various locations on both Lower and Upper Tall el-Hammam. To collect ant sand from other sites in the TMG, however, I had to get a special permit from the Department of Antiquities (DoA), and doing so earned me the title of "Mr. Ant Sand" within the Department. Armed with the permit and accompanied by Mr. Qutaiba Dasouqi, Chief Surveyor for the DoA, I collected more 5-dram vials of ant sand from or near seven other sites (Tall Mweis, Tall Iktanu, Tall Rama, Umm Haddar, Tall Tahuneh, Tall Kafrayn, and Tall Nimrin). I also collected seven 5-dram vials of sand/soil from the Waqf as Suwwan meteoritic impact crater located about 150 km SE of Tall el-Hammam. After returning home, I brought samples (see Figure 33) from each of the 61 vials to New Mexico Tech for microscopic examination.

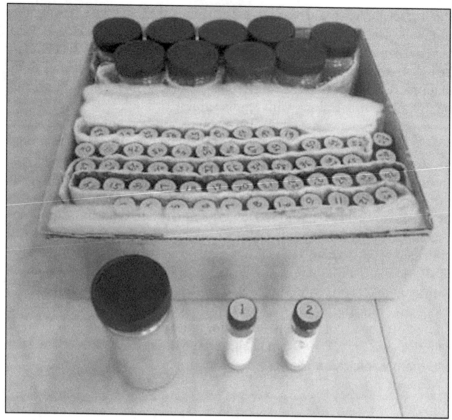

Photo by P. Silvia

Figure 33 - Soil, Ash, and Ant Sand Samples

A deep probe was excavated during TeHEP Season Five (2010) in the Admin Area of Field LS that cut through an ash layer that we believe is associated with the MBA destruction layer. During Season Nine (2014), I removed the 5 mm thick crust of salts that had leached out of the vertical walls of the probe and cored into a section of the south and north walls to remove six samples of soil and ash from each side (see Figure 35). From the south side I collected a set of six samples from the south side in 25 dram vials for my own use. From the north side I filled six quart-sized freezer bags, each weighing about 0.8 kg, and shipped the bulk samples to A. West of Geoscience Consulting in Dewey, AZ, for distribution to the materials analysis team.[42]

---

[42] See Appendix D for a listing of the materials analysis team members.

Figure 34 - Soil/Ash Sample Locations

West sent portions of each sample to fellow researchers at Northern Arizona State University (T. Bunch), Elizabeth City State University in North Carolina (M. LeCompte), and DePaul University (W. Wolbach). All of these researchers are highly experienced in the investigation of meteoritic airburst and impact events. Collectively, they have contributed to a greater understanding of a hypothesized meteoritic impact/airburst event that occurred ca. 12,900 YrBP across four continents that is associated with the Younger Dryas cooling and reorganization of the Clovis people, and the extinction of at least 35 mammal genera in North America (Firestone, *et al*, 2007; Bunch, *et al*, 2012). Their purpose in examining these materials was to see if they contain any of the proxies of a cosmic impact event.[43]

Pottery Samples

Pottery samples were collected from only Tall el-Hammam for my study. Samples of vitrified pottery collected during previous seasons were provided by S. Collins, director of the Tall el-Hammam Excavation

---

[43] Additional samples were collected during Season Ten from the MBA destruction layer in Square UA.7GG, but analysis of these samples is still very preliminary.

Project. Additional samples were collected during Season Nine (2014) specifically for this study.

Over 10,000 pieces of broken pottery are recovered during a typical excavation season at Tall el-Hammam.[44] Of these, only about 10% are "diagnostic" sherds (pieces of rims, handles, bases, spouts, or containing some form of decoration) are kept and registered because the form (shape), material, and firing/hardness of these pieces are indicators of the period in which they were made and are therefore used to help date the context in which they were found.

Thus far, fewer than a dozen pieces of vitrified pottery have been found (see Figure 19, above), and these are the pieces of specific interest to this study. What makes these pieces unique as well as extremely rare is that only one side has been melted into glass rather than the entire piece. Since Bronze Age potters lacked the technological skill to produce this effect, one may reasonably to conclude that the cause of this transformation was not anthropogenic (i.e., a human-caused change).

Several small (1 cm² or less) pieces of peculiar, pumice-like pottery were found during Season Nine which I dubbed "Pumiceware" (see Figure 35). These pieces had distinct tool marks from the potter on one side, but the finished surface on the other side was missing. What should have been reasonably solid clay pottery was black and full of tiny holes that appeared to be lined with greenish glass. I saved these pieces and brought them home for further study. During our final reading of the pottery back in the U.S., I found a larger piece of "normal" pottery, identified as having come from the same excavation square and locus as the aforementioned pieces, which had identical tool marks and color on one side and a red slip coating on the other. The clay of this piece was also black. I therefore concluded that this was a piece of the same "parent vessel" from which the Pumiceware pieces came. After photographing them, I sent them to the researchers at Northern Arizona University for detailed examination along with a sample of vitrified pottery.[45]

---

[44] By contrast, only 82 whole (complete or nearly complete) vessels have been recovered and registered from Tall el-Hammam during nine seasons of excavation. The scarcity of whole vessels is presumed to be related to the violent destruction that hit the site during the MBA.

[45] This sample of vitrified pottery had been examined previously at New Mexico Tech using different technology than that available at Northern Arizona University.

Top Left: Inside, unaltered, showing tool marks.
Top Right: Outside, unaltered, showing original color.
Bottom: "Pumiceware" chip showing glass-lined holes.

Photos by P. Silvia

Figure 35 - "Pumiceware" versus "Normal" Pottery

Rock Samples

Prior to the initiation of this study, the only rocks collected as possible evidence of a cosmic event were the melt rocks found previously by S. Collins at Tall Mweis (see Figure 2, above). My interest in rocks changed, however, when I had opportunity to visit the Waqf as Suwwan meteoritic impact crater on April 4, 2014, with a group from the Friends of Archaeology and History (FOAH) in Amman. This field trip was organized by A. Abu-Shmais, Deputy Director of FOAH, and led by E. Salameh, Professor of Geology at the University of Jordan. Waqf as Suwwan is the only sizable impact structure to be identified anywhere within the entire region of the Middle East. This opportunity presented itself quite unexpectedly while I was in residence at ACOR doing research on the TMG, and it proved to be very instructive on the effects of a meteoritic impact event.

The Waqf as Suwwan meteor crater is located about 150 km southeast of Tall el-Hammam in a remote area of the eastern desert of Jordan, close to the border of Saudi Arabia (see Figure 36). Originally called "Jabel Waqf as Suwwan" (Arabic for "Mountain of the Upright Chert"), the site was misidentified in the 1960s as a crypto-volcanic structure, but the total lack of volcanic rock in, around, or near Waqf as Suwwan prompted a new investigation of the site in 2005. The deeply eroded ring structure of Waqf as Suwwan was subsequently identified as a meteor crater and registered as such in 2006 (Salameh, *et al*, 2008).

A dense network of wadis crossing the structure hints at substantial erosion (see Figure 37). A ring of hills about 6 km in diameter marks the nearly circular outer rim of the crater. An inner ring, nearly 1 km in diameter, rises about 50 m above the surrounding floor of the crater. The oldest exposed rocks of the inner ring are sandstones that have been pushed up from their normal depth of 400-600 meters below the surface. The sandstone is varied in color with reddish, violet, brown and yellowish tones. Around the periphery inside the inner ring are numerous blocks of quartzite. On the macroscale, the strata of the inner structure are extensively deformed and fractured (Salameh, *et al*, 2008).

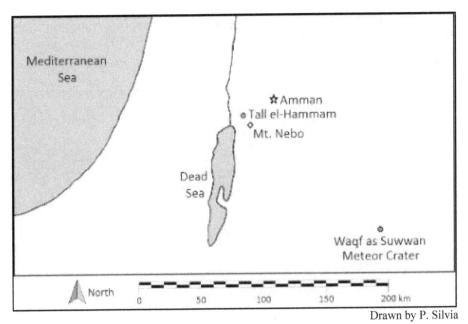

Drawn by P. Silvia

Figure 36 - Waqf as Suwwan Location

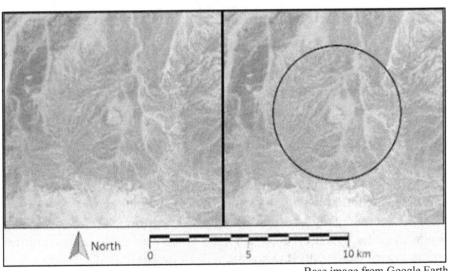

Base image from Google Earth

Figure 37 - Waqf as Suwwam Meteor Crater (Vertical View)

A central mound is present within the inner rim that is presumed to have been formed by a centripetal inflow of material back into the crater immediately following the impact (see Figure 38). Current estimates of the crater formation suggest an initial crater depth of 1.5-1.8 km and an excavation depth 500-600 m below that for the impactor. Approximately 500-700 m of erosional fill has accumulated since the impact to raise the crater floor to match the current levels (Heinrichs, *et al*, 2014).

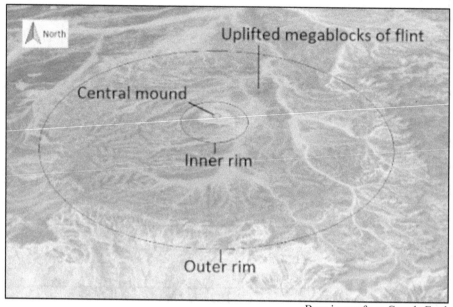

Figure 38 - Waqf as Suwwam Meteor Crater (Oblique View)

Large scale structures and the geometries of overturned folds were carefully mapped and interpreted as indicators of an oblique impact to the NE. Magnetic and seismic investigations of the crater by a team of scientists from the University of Jordan and the Geoscience Center in Goettingen, Germany, concluded that the slight asymmetry of the nearly circular Waqf as Suwwan crater also suggests an oblique impact directed toward the NE. This conclusion was strengthened by the discovery of a debris scatter field outside and to the NE of the outer rim.[46]

The estimated age of the alluvial fan surrounding the Waqf as Suwwan crater is about 100,000 YrBP, which sets a maximum limit for dating the impact event. It is not possible at this time to determine a precise date of the crater formation from the materials in and around the immediate vicinity of the crater, but a team of anthropologists studying

---

[46] Information regarding the debris field and possible dating of the impact event was provided through a private conversation at the American Center for Oriental Research (ACOR) with Prof. Dr. Klaus Bandel from the University of Hamburg, Germany.

the material remains of ancient inhabitants of the region noticed a dramatic change of diet around 8,200 YrBP. They suggested that the impact may have wiped out the people living there at the time, with the people who replaced them coming from a different area with different dietary preferences. Hence, this would suggest an impact date as recent as 8,200 YrBP, or ca. 6200 BCE.

The remote location of the crater has sheltered it from large scale human contamination and pillaging, although some attempt was made to do some unauthorized quarrying of limestone. That turned out to be a good thing, however, because numerous shatter cones were exposed in the fractured blocks of limestone, and these shatter cones were a major reason for Dr. Salameh and his fellow investigators to identify Waqf as Suwwan as an impact crater (see Figure 39).

A. Fractured flint; B. Agglomeration of flint chert bound by meltrock;
C. Uplifted megablocks of flint; D. Limestone shatter cones.

Photos by P. Silvia

Figure 39 - Waqf as Suwwan Impact Materials

The entire site is littered with fine-grained flint chert. Large blocks of fractured flint, many standing on edge, protrude from the rim walls and crater floor. Megablocks of flint measuring several meters across and high protrude from the crater floor about 1 km NE of the central mound and ½ km outside of the central ring structure. Scattered about the crater floor are also many agglomerations of crushed flint bound by either meltrock or by infilling with a groundmass of carbonate and/or barite through hydrothermal activity immediately following the impact. Also to be seen are blocks of marly limestone displaying a cross-thatched fracture pattern indicating impact stress (see Figure 40).

A,B - Crushed flint agglomerations.
C,D - Brittle deformation of marly limestone.

Photos by P. Silvia

Figure 40 - More Waqf as Suwwan Impact Materials

Except for the broken limestone blocks containing shatter cones, all of the aforementioned materials were found (and photographed) *in situ*. All of the materials shown in the previous figures were too large to bring back for study, but I was able to collect and bring back the smaller samples shown in Figure 41, below.

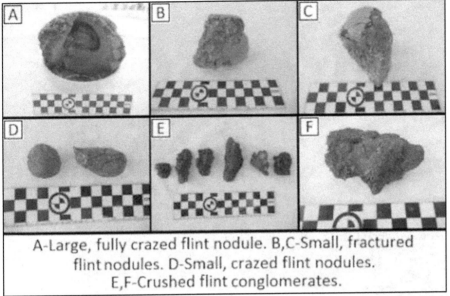

A-Large, fully crazed flint nodule. B,C-Small, fractured flint nodules. D-Small, crazed flint nodules. E,F-Crushed flint conglomerates.

Photos by P. Silvia

Figure 41 - Waqf as Suwwan Samples

The samples collected from Waqf as Suwwan provided great insight into the expected appearance of materials that have been subjected to impact stress. Equipped with this new knowledge, I returned to Tall el-Hammam with my wife Yvonne on April 8, 2014, to search for and collect similar material. Evidence collected during the previous nine excavation seasons suggests that the destruction event that hit Tall el-Hammam during MB2 approached from the SW; therefore, we limited our half-day search to the south-facing slope of the Upper Tall, which will be subsequently discussed.

The surface of Upper Tall el-Hammam is littered with rocks (see Figure 42). Most of these are assumed to be from the bulldozed trench that runs almost the entire length of the Upper Tall (Collins, *et al*, 2006). Consequently, establishing provenance for the surface rocks is virtually impossible. That being said, we were still able to find numerous rocks that exhibited features very similar to those that I saw at Waqf as Suwwan (compare Figure 41, above, with Figure 43, below; also, Figure 44, below).

Photo by M. Luddeni
Figure 42 – West End of Upper Tall el-Hammam

Whereas the primary surface rocks at Waqf as Suwwan are flint and, to a lesser degree, marly limestone, the surface rocks on Upper Tall el-Hammam are mostly marly limestone and sandstone with some flint. The percentage of surface rocks from Upper Tall el-Hammam showing signs of thermal stress was significantly lower (less than 5%) than the percentage of surface rocks showing signs of impact/thermal stress at Waqf as Suwwan (greater than 80%), however. This difference can probably be attributed, to some extent at least, to the undisturbed setting of Waqf as Suwwan, whereas all of the stones at Tall el-Hammam had been relocated numerous times for construction purposes.

The larger rocks at Upper Tall el-Hammam showed the same cross-thatched fracture pattern that I observed and photographed at Waqf as Suwwan (see Figure 43). The surface crazing on the smaller rocks from both sites, which I brought back for analysis, showed similar surface crazing (see Figure 44). The samples that I brought back from Waqf as Suwwan are flint, and the samples from Tall el-Hammam are marly limestone.

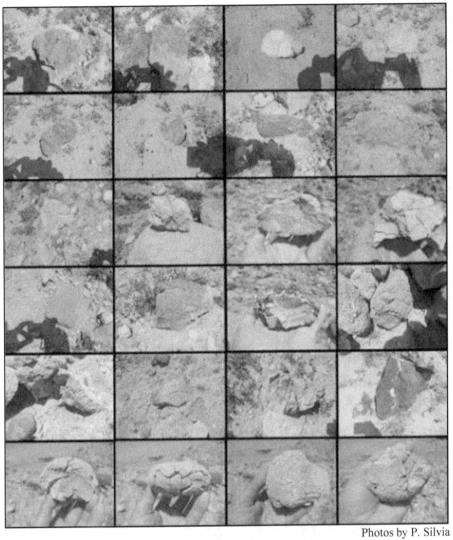

Photos by P. Silvia

Figure 43 - Rock Samples from Upper Tall el-Hammam

91

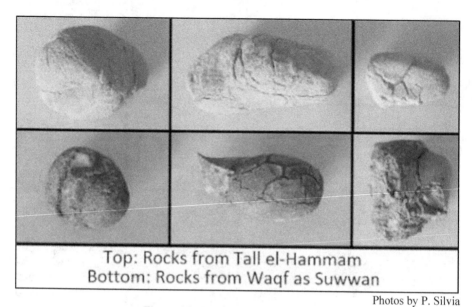

Top: Rocks from Tall el-Hammam
Bottom: Rocks from Waqf as Suwwan

Photos by P. Silvia

Figure 44 - Rock Sample Comparison

Chapter 12. ANALYSIS OF THE SAMPLES

Much of the materials collected from Tall el-Hammam and surrounding sites were analyzed by researchers involved in the investigation of possible meteoritic airburst or impact events ca. 12,800 YrBP (Petaev, *et al*, 2013) that are hypothesized to have triggered the Younger Dryas, a millennium-long cooling period amid postglacial warming that is well documented in the Greenland ice cores (Mangarud, *et al*, 1974). Their collective expertise with examining high-temperature melt products from 18 dated Younger Dryas Boundary (YDB) sites across three continents (North America, Europe, and Asia), spanning 12,000 km around nearly one-third of the planet, proved invaluable to this study.

Pottery Analysis

The original sample of vitrified pottery (see Figure 45) was examined in laboratories at New Mexico Tech in 2006 by N. Dunbar. The primary purpose of this initial examination was to determine if the vitrified[47] surface is actually melted clay and how the observed effect might be accomplished. This analysis concluded that the matrix has basically the same composition from the upper melted region to the lower original matrix and contains 46% SiO_2, 25% CaO, 16% Al_2O_3, 5% FeO, 4% MgO, 2% K_2O, and 2% other materials.[48]

---

[47] Vitrified—turned to glass by melting.
[48] Hand-written notes from N. Dunbar on data sheets that were generated during the analysis.

Photo by P. Silvia

Figure 45 - Vitrified Pottery Cross-section

The vitrified surface itself is very similar in appearance and color to trinitite, the slightly radioactive green glass that was produced from fused soil by the explosion of the first atomic bomb at Trinity Site on the White Sands Missile Range in New Mexico on July 16, 1945 (Eby, *et al*, 2010). The initial assessment of the dynamics required to produce the vitrified surface included a millisecond-duration pulse of heat >8000° C followed by immediate cooling after the heat wave passed. The rationale for this assessment included several factors, the three most important being:

1. The exposure temperature of the sherd had to be at least 8000° C to melt the surface enough that the viscous glass could flow over the broken edge of the sherd.

2. The duration of the heat exposure had to be brief because internal thermal discoloration penetrated only about half of the 5 mm thickness of the sherd.

3. The duration of the heat exposure had to be *extremely* brief (1 millisecond or less) because only the surface of the sherd was melted, and the nominal depth of melting is less than 1 mm.

The glass of the surface was found to be calcium-rich, with little crystal growth other than some plagioclase[49] and pyroxene.[50] This suggests a uniform melting of the clay matrix and mingling of the various mineral components. Quartz crystals embedded in the glass remained relatively intact, other than some minor surface melting (see Figures 46 & 47).

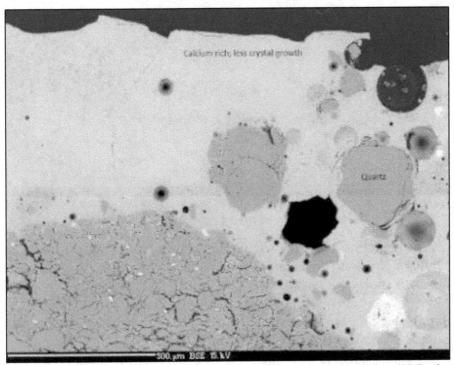

Image from D. Burleigh and N. Dunbar

Figure 46 - Vitrified Surface Cross-section, 500 um Scale

---

[49] Plagioclase—a feldspar consisting of sodium and calcium aluminum silicates.
[50] Pyroxene—a group of dark green, brown, or black silicate minerals containing varying amounts of calcium, aluminum, iron, magnesium, and sodium.

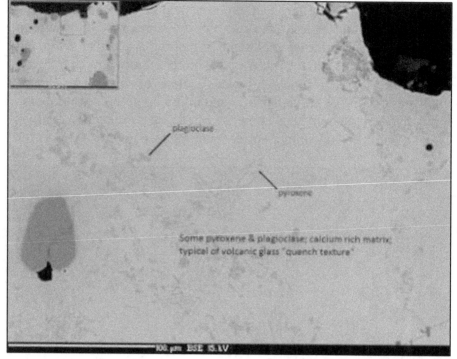

Image from D. Burleigh and N. Dunbar
Figure 47 - Vitrified Surface Cross-section, 100 um Scale

A re-examination of the vitrified sherd by T. Bunch and J. Wittke at Northern Arizona University (NAU) in 2014 produced similar conclusions. Plagioclase crystals start to form as the temperature drops to ~1300° C, and this is followed by the formation of pyroxene crystals. Both plagioclase and pyroxene crystals appear in the sherd, so that sets a lower bound for the melt temperature.[51]

A third examination of the vitrified pottery sherd was conducted in early 2015 by M. LeCompte at Elizabeth City State University and A. V. Adedeji at North Carolina State University (NCSU). Using Scanning Electron Microscopy (SEM), a zirconium enriched nugget was found embedded in the pottery host matrix just under the glassy surface (see Figure 48). Based upon the presence of vesicles (tiny pin holes) that appear on portions of the nugget in zoom images, it appears that the nugget reached a boiling temperature of approximately 4000° C, hotter than the

---

[51] Personal correspondence with T. Bunch, principal researcher for this second look at the vitrified pottery and first look at the Tall Mweis melt rock.

surface of the sun at points within the sun's chromosphere inside a sunspot.[52] A chromium nugget was also found near the glassy surface that contained a few vesicles, but they are much harder to see because of their smaller size.

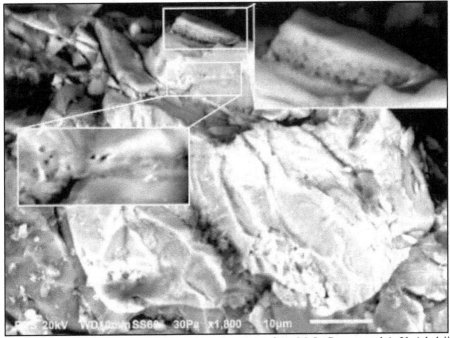

Image from M. LeCompte and A. V. Adedeji
Figure 48 - Zirconium enriched nugget in vitrified pottery sherd

The temperatures required to produce the observed effects in the vitrified pottery sherd are inconsistent with pottery making technology of the Early Bronze Age, where temperatures were typically between 480° C (900° F) and 650° C (1,200° F). Although vitrified silica-based-based materials are occasionally found inside architectural settings where suitable fuels and updrafts create temperatures in excess of 1,100° C (2000° F), this vitrified sherd was found in a thick destruction matrix of ash, churned up with an array of disintegrated cultural remains. It was not found in an architectural context, such as under a collapsed roof. Thus, it probably experienced the high-temp melting of its surface before it was buried in the debris matrix. It was likely a storage jar that was sitting

---

[52] Unpublished preliminary report received through personal correspondence from M. LeCompte.

exposed on a rooftop or open plaza, with the shoulder of the vessel from whence the sherd came having maximum exposure to the heat source. The force of the thermal blast likely caused the vessel to shatter at the very moment the melted surface was already solidifying, just barely lapping over the edge of the break. The blast churned it up with other materials and ash, which subsequently landed in the position where it was found during excavation of Field UB at Tall el-Hammam.[53]

The internal temperature of the melted glass is established with a reasonable degree of accuracy through the mineral (quartz) and crystalline (plagioclase and pyroxene) content. What has not yet been determined with certainty is the combination of the actual temperature to which the sherd was exposed and the duration of that exposure. The conclusion of the NMT examination proposed a temperature in excess of 8,000° C for less than a millisecond. The NAU researchers required further analysis before offering a temperature/duration profile, but they conceded the temperature gradient visible in the sherd goes from > 1,700° C at the surface where melting occurred to little evidence of secondary melting below a few mm of the surface, which strongly suggests that the melting event was a flash exposure to intense heat.[54]

The most recent (February 2015) images from NCSU show a fractured quartz grain embedded in the glazed surface that appears to have melted and flowed into the surrounding clay matrix as indicated by the pollywog tail of the grain curving faintly to the lower left in the SEM image (see Figure 49). A second image (see Figure 50) shows a zircon grain that also appears to have both fragmented and melted into its surrounding matrix as revealed by the fading bright material on the inside of the "C" of the Zirconium enriched grain. Both of the grains shown in these two images require extremely high temperatures to become molten.[55]

[53] Personal correspondence with S. Collins.
[54] See Bunch, *et al*, 2012 for mineral melt temperature references.
[55] Analysis of the images in Figures 49 and 50 was provided by M. LeCompte through personal correspondence.

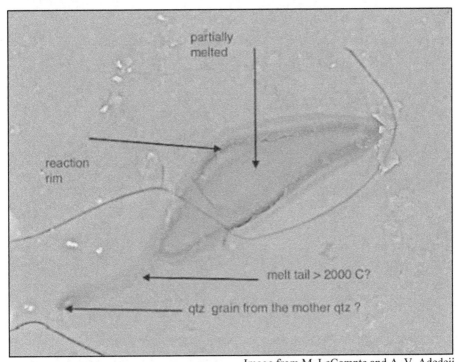

partially
melted

reaction
rim

melt tail > 2000 C?

qtz grain from the mother qtz ?

Image from M. LeCompte and A. V. Adedeji
Figure 49 – Melted, fractured quartz grain with pollywog tail

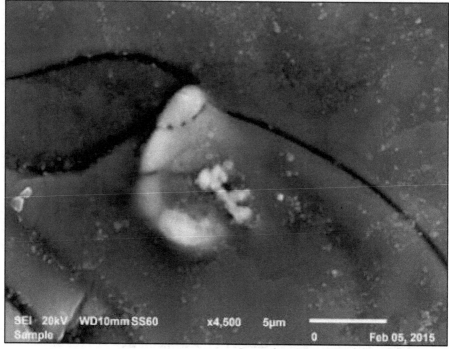

Image from M. LeCompte and A. V. Adedeji
Figure 50 - Fractured and melted Zircon grain

Considering the very shallow depth of the glassified surface, the exposure temperature required to produce the melted minerals observed in the vitrified pottery sample is now estimated at 2-3 times the maximum melt temperature, or between 8,000 and 12,000° C with an exposure time of no more than a few milliseconds. These estimates will be further refined through continued testing beyond what has been conducted thus far for this dissertation.

Melt Rock Analysis

The melt rock recovered from Tall Mweis was examined by T. Bunch from NAU.[56] He identified three different lithologies[57] in the melt

---

[56] The material presented in this section was provided by T. Bunch through personal correspondence and has not yet been published.

[57] Lithology—composition and characteristics of a rock.

rock. Each lithology is a conglomerate of rounded clasts[58] of quartz, sandstone, quartzite, gypsum, carbonates, and other materials that are common to the region. It is possible that the individual lithologies were formed through separate cycles of erosion, deposition, and induration[59] to form typical sedimentary rock formations. Another possibility is that the separate lithologies were welded together through a flash melting process.

The main mass of lithology A appears to be fused, but not melted, orthoquartzite sandstone fragments (see Figure 51). The surface of Lithology A is fully melted and covered by a SiO_2 glass coating. The high points are thinly coated with glass, but the depressions are more thickly coated as if the glass flowed into these low spots. The glass is also a bit thicker near the edges of the top surface as if it flowed toward and wrapped over the edge to the bottom side.

---

[58] Clast (n.)/clastic (adj.)—a type of sedimentary rock (such as shale, siltstone, sandstone or conglomerate) or sediment (such as mud, silt, sand, or pebbles). Clastic rocks are accumulations of transported weathering debris that have been lithified.

[59] Induration—the process of solidifying or hardening a material into a rock through pressure, cementation or heat.

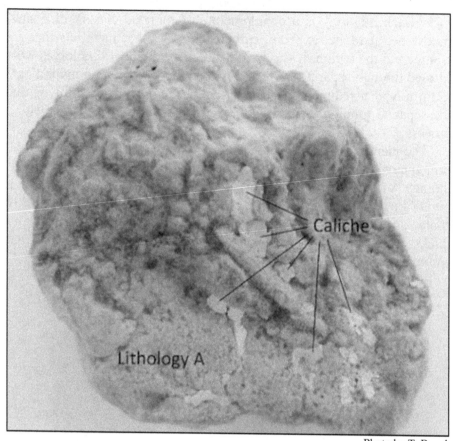

Photo by T. Bunch
Figure 51 - Melt Rock Top View

Lithology B is composed of a mixture of gypsum and sandstone (see Figure 52). The vesicles[60] in lithology B are deep and tube-like with only a few that are more bubble-like.

---

[60] Vesicle—spherical or elongated cavities in an igneous rock that are created when a melt crystallizes with bubbles of gas trapped inside.

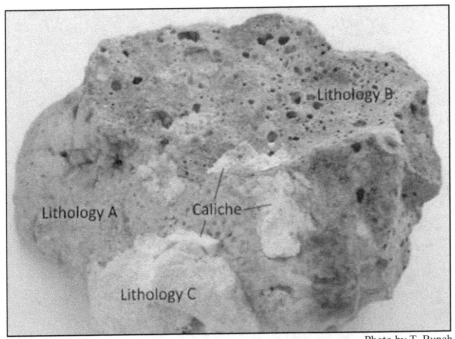

Figure 52 - Melt Rock Bottom View

Photo by T. Bunch

If the presence of gypsum is correct, then these strange vesicles may be due to out-gassing of H_2O and SO gases during the rapid thermal decomposition of gypsum and carbonates, in addition to pore water. Gypsum and carbonates only melt under very high temperatures, and rapid melting requires temperatures greater than 1,500° C for less than a few seconds (Jones, *et al*, 2013).

Lithology C (see Figure 53) is a fused, but not melted, clump of sandstone that appears to have plopped into the still molten surface glass of lithology A with partial melting of the margins. When viewed from the edge, the bluish melt of lithology A clearly appears to have been displaced while molten by the collision of lithology C. Partially melted grains of lithology C are visible in the glass of lithology A as a result of this low velocity impact accretion. The welding of lithology C to the melt of lithology A implies a glass temperature greater than 1800° C during accretion (Bestmann, *et al*, 2011).

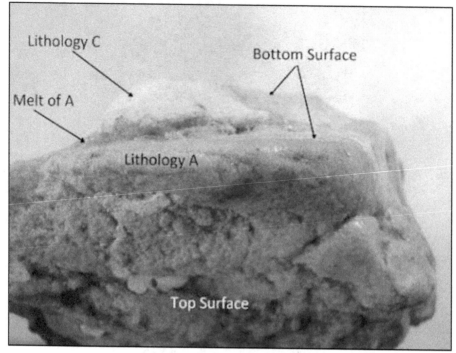

Lithology C

Melt of A

Bottom Surface

Lithology A

Top Surface

Photo by T. Bunch

Figure 53 - Melt Rock Edge View

A thin-to-thick melt veneer covers the entire rock (hence the term "melt rock" to refer to the entire mass), except for a portion of lithology B. The glass is very clear, has a hardness of 6.5 on the Moh scale, and is mostly SiO_2. The glass also contains a lot of Al_2O_3 and CaO, which is consistent with the area geology. The complete glass composition is shown in Table 3, below.[61]

---

[61] Mineral analysis of the melt glass was performed by T. Bunch and communicated by private correspondence.

Table 2 - Melt Rock Glass Composition

SiO_2	62.02
Al_2O_3	13.95
CaO	10.09
MgO	4.98
FeO	3.68
K_2O	2.79
Na_2O	0.64
TiO_2	0.62
SO_3	0.51
MnO	0.32
P_2O_5	0.23
Cr_2O_3	0.04

Some areas of the glass coating are bluish-to-greenish in color (see Figure 54), possibly due to reducing conditions during melting and the presence of red ferric iron and a tiny amount of Fe_2O_3 (<0.5% by weight). This only happens under very low oxygen fugacity[62] and is extremely rare under typical geogenic (naturally occurring; e.g., volcanic) conditions, but is common in trinitite, fulgurites[63], and YDB melt rocks (Bunch, *et al*, 2012), i.e., flash heating/melting events. Pieces of caliche[64] are clearly visible on all surfaces of the melt rock. Because caliche has a much higher melting temperature than SiO_2, the melt pulls away from the caliche so that it appears to "float" in the glass. Trapped air bubbles are visible in some of the thicker areas of the glass veneer (see Figure 55.)

---

[62] Frugacity—the amount of O_2 available during the melt.
[63] Fulgurite—glass, typically in the form of tubes or beads, created by lightning when it strikes sand.
[64] Caliche—a sedimentary rock consisting of calcium carbonate that forms a hardened natural cement.

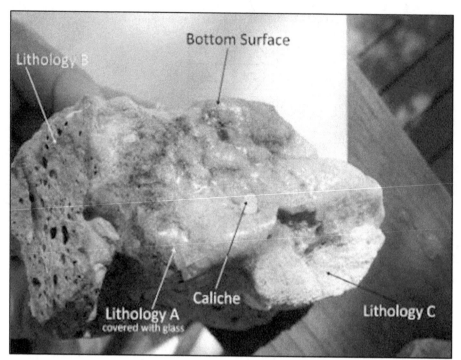

Figure 54 - Melt Rock Edge View from the Other End

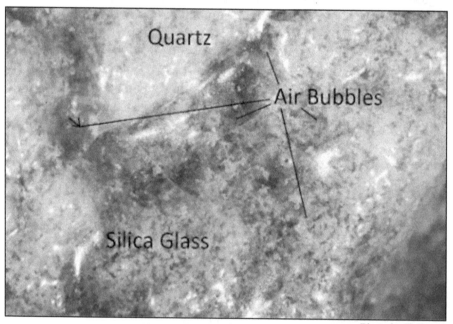

Figure 55 - Melt Rock Glass Surface, x10

106

A close look at the interior of lithologies A and B was achieved by cutting across the tapered edges of the melt rock (see Figure 56). Lithology A is coated with SiO_2 glass, below which is mostly pure quartz sandstone that shows pockets of SiO_2 glass and interstitial melt zones (see Figure 57). Vaporization bubbles are small but common.

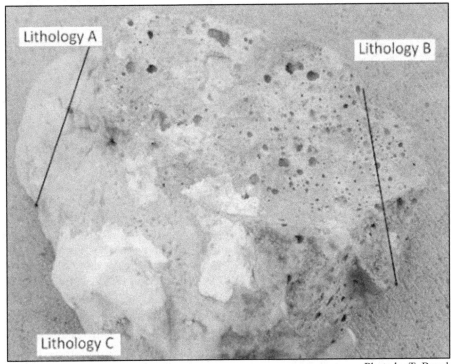

Photo by T. Bunch

Figure 56 - Melt Rock Cross-section Cut Lines

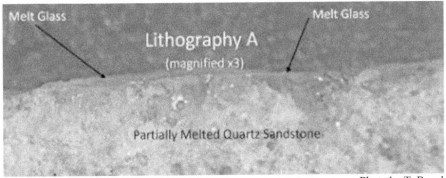

Photo by T. Bunch

Figure 57 - Lithography A Cross-section

107

Lithology B is actually a composite of several different components that include quartz-rich sandstone, gypsum, and a few that are unidentified, all of which either completely melted followed by rapid crystallization of quench crystals (particularly common for previous gypsum rock components) or show various degrees of melting into clear glass (mostly quartz-rich components). Voids (irregular bubbles) within lithology B are lined with chocolate-brown glass (see Figure 58).

Photo by T. Bunch

Figure 58 - Lithography B Cross-section

Most of lithology B lacks the outer melt glass coating, especially on the relatively flat bottom side. Lithology B also has a "fractured" appearance with crisp edges that suggests that a piece, or pieces, may have broken off [Bunch]. Collins, however, reported that the melt rock was extracted from a surface sand matrix next to a stone wall foundation. It was not connected to, or related to, a "mother" rock. Although the origin of the "mold depression" in lithology B in unknown, it could originally have been slammed against a stone or mudbrick when it was still viscous and later separated after cooling.

Lithology B is very similar to large (8-10 cm) melted rock composites found scattered over 1600 sq km of the Dakhleh site in southwestern Egypt that is considered to be a product of an aerial burst ~ 200,000 years ago (see Figure 59). Apparently, these melted composites were made from melting of the surface sediment composed of fine to pebble-size materials.

Figure 59 - Dakhleh Site Melt Rock

Quartz melts at >1720° C and boils at >2230° C (Bunch, *et al*, 2012; Bestmann, *et al*, 2011). Some of the observed bubbling in the melt rock may be due to rapid vaporization of water, however. Gypsum does not normally melt, but thermally decomposes at temperatures >1500° C, unless the temperature is very high and applied rapidly for less than a few seconds, as would be typical of aerial burst conditions. Examination of the melt rock under the SEM also revealed melted clasts and zircon grains internal to the melt rock (see Figure 60). Both of these melt around 2000° C (Fel'dman, *et al*, 2006; Gueguen, *et al*, 2014).

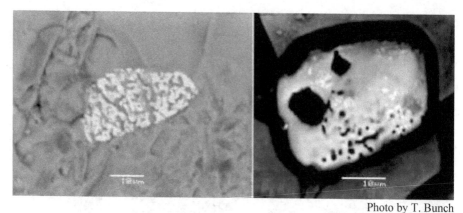

Photo by T. Bunch

Figure 60 - Melted clast and zircon in the melt rock

The temperatures required to produce the melt rock are higher than are natural for geologic mechanisms and any ancient industrial process (Bestmann, *et al*, 2011). A similar result could be achieved in a modern electric arc furnace, but the melt rock is too "clean" (totally lacking contamination clinging to the glass surface). All things considered, the melt rock is very dissimilar to typical industrial products and was found in an isolated archaeological context, albeit on the surface of the ground.

The water content of amorphous[65] anthropogenic (man-made) and geogenic glasses is typically 1,000s ppm (Le Losq, *et al*, 2012), whereas similar samples of hard impact, fulgarites, YDB, and trinitite glasses all have extremely low water content, ~400 ppm (Bunch, *et al*, 2012). This difference is attributed to the extreme melting/heating temperature encountered by the latter group that vaporizes the water content and partially accounts for the large and numerous vesicles from rapid outgassing.

Anthropogenic and geogenic glasses are susceptible to devitrification[66] through weathering processes, but the melt rock looks like it was formed yesterday. The general features of the melt rock are consistent with production by a thermal blast produced by a meteoritic airburst or a nearby airburst (perhaps at the Dead Sea?). The textures of the melt rock imply tumbling of the three separate lithologies in the turbulent blast, followed by hot, ballistic accretion of these objects with continued melting to form the glass veneer. For such an old rock, even from a desert

---

[65] Amorphous—a non-crystalline solid, such as glass, in which the atoms and molecules are not organized in a definite lattice pattern.

[66] Devitrification—the conversion of the amorphous glass structure into fine-grained, organized, crystalline material.

environment, it displays very little weathering effects from internal water, although there is some minor external weathering of the gypsum surface and filling of some cavities by caliche.

Finally, the surface of the melt rock is nearly pristine and shows no evidence of fluvial or mass-wasting transport. It could have been brought into the location in which it was found by humans, since it is a "pretty" rock. It certainly was not used as a tool. If the melt rock is the product of an aerial burst, it was most likely formed close to its discovery location, Tall Mweis.

Soil, Ash, and Sand Analysis

Preliminary examination of ash and soil samples that I collected from Tall el-Hammam was conducted by A. West and T. Bunch. West provided a sample mount (stub) with two adhesive areas covered with grains to M. LeCompte and his team at Elizabeth City State University (ECSU) and North Carolina State University (NCSU) in Raleigh for scanning electron microscopy (SEM) examination. Two sessions of SEM examination were performed at NCSU, one on May 2, 2014, and the other on May 23, 2014, with C. Mooney, D. Batchelor, and R. Garcia. Other sessions were performed at ECSU with V. Adedgi on June 27, July 14, 25, 28, 29, and August 1, 2014, and on March 3, 2015.

Initially, LeCompte[67] faced a significant challenge detecting the small spherules (10-20 μm) among the large quantity of larger debris since spherules he had previously examined from other impact events were considerably larger (80-100 μm). Experience obtained during the first session made the second session more productive. About two dozen spherule candidates warranting further study were identified, but more and better imagery is needed to confirm them as impact melt products.

At least one spherule appears to possess the same sort of dendritic surface texturing LeCompte has seen in published literature associated with the Tunguska event and other samples collected from sites associated with the hypothesized YDB impact event. The surface texturing observed on such spherules is indicative of high temperatures and pressures characteristic of a meteoritic impact or airburst. The energetic airburst and/or ground impact vaporizes target rock, and the melted droplets are

---

[67] The electron microscopy analysis information presented in this section was provided by M. LeCompte, assisted by associates at ECSU and NCSU, in an unpublished preliminary report received through personal correspondence.

rapidly quenched in ambient conditions, thus creating the patterns of partial crystallization observed on their surfaces. Such meteoritic impact spherules are typically rich in iron or iron oxides and have small amounts of other elements including aluminum and silicon (Firestone, *et al*, 2007; Bunch, *et al*, 2012; LeCompte, *et al*, 2012). The spherule in Figure 61, below, which was found in an ash-soil matrix from the Tall el-Hammam MBA destruction layer, has that geochemical composition, but also is depleted in oxygen, consistent with high-temperature melting.

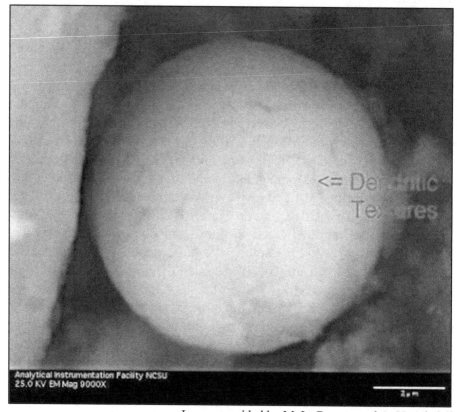

Image provided by M. LeCompte and A. V. Adedeji
Figure 61 - Iron-enriched Magnetic Spherule from TeH

West[68] sent portions of the six soil/ash samples that I had collected by transecting the MBA destruction layer in square LS.42K at Tall el-

---

[68] The geochemical analysis information presented in this section was provided by A. West in an unpublished preliminary report received through personal correspondence

Hammam to Activation Laboratories Ltd. (Actlabs) in Ancaster, Ontario, Canada, for geochemical analysis. The results of this analysis are shown in Figure 62, below. The salt plus sulfate levels in these samples were found to be very high, with a peak (~60,000 ppm, total) occurring in the ash (destruction) layer (sample #4). The significance of these levels is that they exceed the threshold at which the salt concentration becomes toxic to most plants grown for human consumption. According to the USDA, the toxic level of salts for wheat is about 1.3% (12,800 ppm) and for barley is about 1.8% (17,900 ppm).[69] The peak total salt content shown in Figure is 62 is 5.5% (55,000 ppm)!

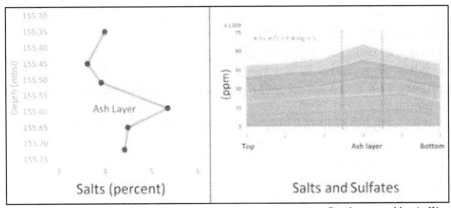

Figure 62 - Salt and Sulphate Content by Sample (% and ppm)

The destruction layer was also found to contain a platinum (Pt) concentration that is about six times higher than the background levels (see Figure 63). This is very similar to data acquired from two YDB sites, one in Syria (Abu Hureyra) and the other in Arizona (Murray Springs), both of which show sharp Pt peaks. West also prepared a "mix" sample by combining material from six vials of soil/sand that I had collected from other locations across Tall el-Hammam and had the "mix" tested for comparison. The Pt level in the "mix" approximately equaled the average for the other five samples from the LS.42K location.

---

[69] USDA, 2011.

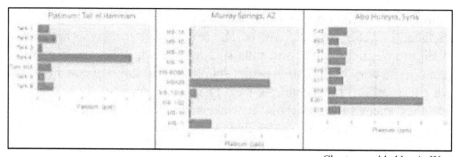

Charts provided by A. West

Figure 63 - Platinum Level Comparison

The tests that West had made on the samples did not detect iridium (Ir), but he suspects that the threshold for this test was set too high. Also, no nanodiamonds were found in the samples that were examined by W. Wolback at DePaul University. West noted, however, that nanodiamonds are not always produced in a meteoritic impact or airburst event, so their presence is not crucial to making a case for such an event at Tall el-Hammam.

I personally examined the "ant sand" samples that I collected from Tall el-Hammam and other sites surrounding Tall el-Hammam, but the results of that examination have thus far been inconclusive. The magnification power of the optical microscope I used for this examination was insufficient to clearly identify any spherules of the size found in the ash-soil matrix by LeCompte using the SEM. Portions of these samples were also examined by students at New Mexico Tech under the direction of D. Burleigh. Some possible spherules were observed in the samples, but they were unsuccessful in isolating them for SEM examination.

The grains of sand are very "dirty" in that they are coated with a much finer grit and very fine clay silt, all bound together by gypsum. In order to see the larger grains more clearly (see Figure 64), I used an ultrasonic cleaner to dissolve the gypsum and "wash" small samples of the ant sand in a vial with demineralized water. I then used a paper filter to separate the sand and fine grit from the water and clay. After drying the filter on a small hot plate, I used a double-sided adhesive strip to pick up the sand and grit from the filter and mount it on white card stock for viewing under the microscope.

114

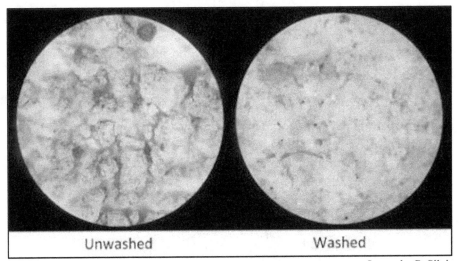

Unwashed   Washed

Image by P. Silvia

Figure 64 - Ant Sand, x200

About 80% of the washed sand is quartz grains. Most of the grains are relatively smooth and semi-rounded, the apparent result of erosional tumbling. I was unable to discern any obvious melt products through my optical microscope. None of the ant sand samples were comingled with obvious ash, however. Therefore, it is probable that the ants had not tunneled into or through a destruction layer.

Chapter 13. AIRBURST PHENOMENOLOGY

The difference between a meteoroid and an asteroid is a matter of size, but there is no formal limit on the size of the biggest meteoroid or the smallest asteroid, however, anything bigger than 10 m in diameter is usually considered an asteroid (Cooke, 2013). About 90% of meteoroids in the inner Solar System come from comets. Most of the asteroids reside in the region of our solar system between the orbits of Mars and Jupiter known as the asteroid belt (see Figure 65). The main belt spans a distance 2.12 and 3.3 AU[70] from the Sun.[71]

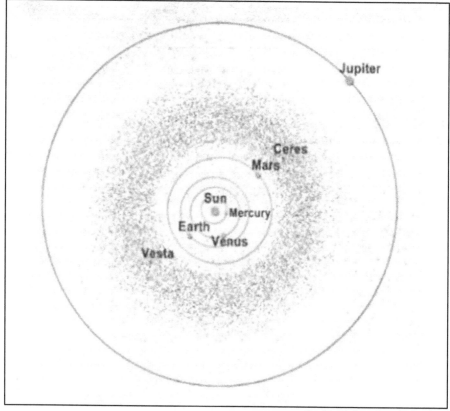

Credit: University of Maryland

Figure 65 - The Asteroid Belt

---

[70] Astronomical Unit (AU) is the average distance from the Sun to the Earth, about 150 million km (93 million miles).
[71] http://astronomy.swin.edu.au/cms/astro/cosmos/A/Asteroid+Belt

Asteroids range in size from small rocks to the largest body, Ceres[72], which was discovered in 1801 by Giuseppe Piazzi and is a little less than one-third the diameter of Earth's Moon. Ceres is actually massive enough that its gravity has pulled it into an approximate spherical (oblate spheroid) shape with a polar diameter of 909 km and an equatorial diameter of 975 km. This shape has resulted in its being classified as a dwarf planet. Vesta[73] is the third largest asteroid and is irregular in shape, measuring about 578 by 560 by 458 km. One of the most interesting features of Vesta is a giant crater near its south pole that measures 460 km across and 13 km deep. The massive collision that created this crater gouged out about one percent of the asteroid's volume, blasting over 1.5 million cubic km of rock into space. The resultant debris, ranging in size from sand and gravel to boulder and mountain, was ejected into space where it began its own journey through the solar system. Scientists believe that about 5% of all meteorites we find on Earth are a result of this single ancient crash in deep space.

There are an estimated 1.1-1.9 million asteroids in the main belt that are larger than 1 km in diameter and many millions of smaller ones. More than 150 asteroids are known to have a small companion moon (some have two moons). There are also binary (double) asteroids, as well as triple asteroid systems, in which two (or three) rocky bodies of roughly equal size orbit each other as together they orbit the Sun.[74]

In addition to the main belt of asteroids, there are other groups of asteroids that are of interest: Trojans and near-Earth asteroids. Trojans are asteroids that share an orbit with a planet, but do not collide with it because they gather around two special places in the orbit (called the L4 and L5 Lagrangian points). There, the gravitational pull from the sun and the planet are balanced by a Trojan's tendency to otherwise fly out of the orbit. The Jupiter Trojans form the most significant population of Trojan asteroids. It is thought that they are as numerous as the asteroids in the asteroid belt. There are also Mars and Neptune Trojans, and NASA announced the discovery of an Earth Trojan in 2011. Near-Earth asteroids

---

[72] The official astronomical name is 1 Ceres because it was the first of the asteroids to be discovered (http://www.nasa.gov/mission_pages/dawn/ceresvesta/index.html#.VInd_ns8HhU).

[73] The official astronomical name is 4 Vesta because it was the fourth of the asteroids to be discovered. (http://www.nasa.gov/mission_pages/dawn/veresvesta_/index.html)

[74] http://www.nasa.gov/spaceimages/details.php!id=PIA12134

have orbits that pass close by that of Earth. Asteroids that actually cross Earth's orbital path are known as Earth-crossers. As of June 19, 2013, 10,003 near-Earth asteroids are known and the number over 1 kilometer in diameter is thought to be 861, with 1409 classified as potentially hazardous asteroids—those that could pose a threat to Earth—including Earth-crossers (www.nasa.gov).

Jupiter's massive gravity and occasional close encounters with Mars or another object change the asteroids' orbits, knocking them out of the main belt and hurling them both toward and away from the Sun and across the orbits of the planets. Stray asteroids and asteroid fragments have slammed into Earth and the other planets in the past, playing a major role in altering the geological history of the planets and in the evolution of life on Earth. Scientists continuously monitor Earth-crossing asteroids and near-Earth asteroids that may pose an impact danger.

Every day, Earth is bombarded with more than 100 tons of dust and sand-sized particles. The vast majority of these particles are so tiny that they burn up in the atmosphere unnoticed. Roughly 25 times each year a piece of space rock will hit the Earth's atmosphere that is large enough to be called a bolide because of its fiery streak across the sky as it burns up in the atmosphere. About once a year, an automobile-sized bolide streaks through Earth's atmosphere and creates a really impressive fireball, usually with an explosive finale, before reaching the surface. Every 500-2,000 years or so, an asteroid the size of a football field hits Earth and causes significant damage to the area. It is currently believed that an asteroid 1-2 km in diameter striking the Earth would be sufficient to generate worldwide effects. The largest known potentially hazardous asteroid is Toutatis, which measures about 5.4 km in diameter (www.nasa.gov). During the 20-year period between 1994 and 2013, over 550 bolide events were observed worldwide (see Figure 66).

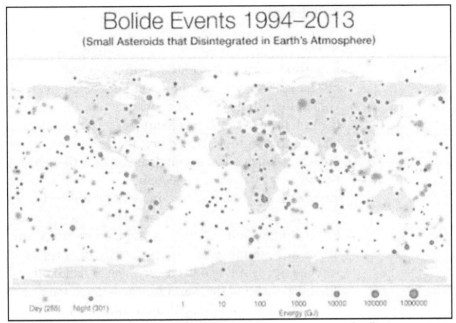

Credit: www.nasa.gov

Figure 66 - Bolide Events of the Past 20 Years

Airburst (bolide) events were happening long before NASA started recording and reporting them, however, as demonstrated below:

Event Date   Eyewitness Description of Event

Sep 20, 1676   In the dusk of the evening…there appeared a sudden light, equal that of noon-day, so that the smallest pin or straw might be seen lying on the ground. And, above in the air [over England], was seen (at no great distance than was supposed) a long appearance as of fire—like a long arm—with a great knob at the end of it; shooting along very swiftly; and, at its disappearing, seemed to break up into small sparks or parcels of fire, like as rockets and such artificial fireworks in the air are wont to do. (*Philosophical Transactions (1665-1678)*, Vol. 12 (1677-1678), pp. 864-866)

Mar 19, 1718   This wonderful luminous meteor was seen in the heavens [over England]…as it was a matter of surprise and astonishment to the vulgar spectator, so it

119

afforded no less subject of enquiry and entertainment to the speculative and curious in physical things. Some of its phenomena being exceedingly hard to account for, according to notions hitherto received by our naturalists, such as the very great height thereof above the Earth; the vast quantity of the matter thereof; the extravagant velocity wherewith it moved; and the prodigious explosions thereof heard at so great a distance, whose sound, attended with the very sensible tremor of the subject air, was certainly propagated through the medium incredibly rare and next to a vacuum. . . .Of several accidents that were reported to have attended its passage, many were the effect of pure fancy[75]; such as hearing its hiss as it went along, as if it had been very near at hand; others imagined they felt the warmth of its beams; and some there were that thought, at least wrote, that they were scalded by it. But what is certain, and no way to be disputed, is the wonderful noise that followed its explosion. All accounts from Devon and Cornwall and the neighboring counties were unanimous, that there was heard there, as it were the report of a very great cannon, or rather of a broadside, at some distance, which was soon followed by a rattling noise, as if many small arms were being promiscuously discharged. What was peculiar to this sound was, that it was attended by an uncommon tremor of the air, and everywhere in those counties, very sensibly shook the glass windows and doors of the houses, according to some, even the houses themselves, beyond the usual effect of cannon. (*Philosophical Transactions (1683-1775)*, Vol. 30 (1717-1719), pp. 978-990)

Nov 26, 1758   This meteor seems to have been vertical over Cambridge [England], or nearly so, and to have taken fire

---

[75] Effects of the meteor's passage reported by the eye witnesses were considered to be "pure fancy" by the writer of the article, but have since been proven to be very real, indeed.

about the zenith of that place, or at least to have appeared first there is a state of ignition. From thence it proceeded directly, almost NNW over several counties in England...Ireland...and Scotland...but soon after becoming vertical to [Lanerk]...part of the tail seemed to break off, and to disperse in bright sparks of fire; whilst the head, into which the remainder of the tail instantly collected, moved on in the same direction. [The account goes on to describe the meteor as reigniting after traveling another 400 miles northward before falling into the sea.] (*Philosophical Transactions (1683-1775)*, Vol. 51 (1759-1760), pp. 259-274)

May 10, 1760   The weather then [9:35 AM] being fair and calm, the people at Bridgewater [Massachusetts], and the towns near it, about 25 miles south from hence, were surprised by a noise, like the report of a cannon, or volley of small arms, which seemed to come from the west. This report was followed by a rumbling noise, which most took for the roar of an earthquake; and, when it had lasted about a minute, there was another explosion. Like that of a cannon; and about as long after, a third; the roaring noise, in the meantime, increasing, so as to fill the all around. . . . After the third explosion, the noise gradually abated, seeming to go off toward the southeast. . . . From a vessel about a league southwest from Cape Cod, and from Martha's Vineyard, we received accounts of a bright ball in the heavens, sufficient to ascertain the reality of the meteor. (*Philosophical Transactions (1683-1775)*, Vol. 52 (1761-1762), pp. 6-16)

Feb 10, 1772   ...exactly at seven in the evening, as I was riding through Tweedmouth [England]...I observed that the atmosphere was suddenly illuminated in a very extraordinary manner. The light of the Moon, which was about half full, seemed to be extinguished by the blaze. . . . I turned round to see from whence the light

121

proceeded, when I beheld a long, bright flame, moving almost horizontally along the heavens. It was conical in form...[and] the base of the cone was rounded like a sphere; and apparently of about one third of the diameter of the Moon at her greatest height. . . . In about ten of twelve seconds it seemed to burst, dividing into a number of small luminous bodies, like the stars of a sky rocket, which immediately disappeared. . . . I had the presence of mind to pull out my watch...to measure the time the report should take in reaching me. . . . I was stunned by a loud and heavy explosion, resembling the discharge of a large mortar, at no great distance, and followed by a kind of rumbling noise, like that of thunder. I examined my watch, and found, that the sound had taken five minutes, and about seven seconds, to reach me; which, according to the common computation of 1142 feet in a second, amounts to a distance of at least 66 miles. (*Philosophical Transactions (1683-1775)*, Vol. 63 (1773-1774), pp. 163-170)

Oct 13, 1838   A meteor exploded...in the Cold Bokkeveld, Cape of Good Hope, with a noise so loud as to be heard over an area of more than seventy miles in diameter, in broad daylight, about half-past nine in the morning. . . . The explosion was accompanied with a noise like that from artillery, followed by the fall of small pieces of matter...portions of which fell or were dispersed on the ground at a distance of...[about] five miles. . . . one piece made a hole as broad as three feet, and sunk deep. It is stated to have been so soft as to admit of being cut with a knife where it first fell; then it hardened... (*Philosophical Transactions of the Royal Society of London*, Vol. 129 (1839), pp. 83-87)

Dec 12, 1882   [Observed by Captain Belknap of the USS Alaska while at sea] ...a few minutes after sunset...a remarkable phenomenon was witnessed in the western horizon from the deck of this ship. . . . The sun had set clear...while the new moon, three days old, gave out a peculiar red light of singular brilliancy. Suddenly, at three minutes before five o'clock, a loud rushing noise was heard, like that of a rocket descending from the zenith with tremendous force and velocity. It was a meteor, of course; and when within some 10° of the horizon it exploded with great noise and flame, the glowing fragments streaming down into the sea like sparks and sprays of fire. (*Science*, Vol. 1, No. 1 (Feb 9, 1883), pp. 4-6)

Jun 30 1908   [From the Itkutsk newspaper, two days after the event]...the peasants saw a body shining very brightly (too bright for the naked eye) with a bluish-white light. . . . The body was in the form of 'a pipe', i.e. cylindrical. The sky was cloudless, except that low down on the horizon, in the direction in which this glowing body was observed, a small dark cloud was noticed. It was hot and dry and when the shining body approached the ground (which was covered with forest at this point) it seemed to be pulverized, and in its place a loud crash, not like thunder, but as if from the fall of large stones or from gunfire was heard. All the buildings shook and at the same time a forked tongue of flames broke through the cloud. All the inhabitants of the village ran out into the street in panic. The old women wept, everyone thought that the end of the world was approaching (Kridec, E.L. 1966. *Giant Meteorites*. Pergamon Press, Oxford)

Aug 13, 1930   ...the sun began to rise on the morning of 13 August 1930 like any other morning...at about eight o'clock, the sun became blood-red and a darkness fell over the region. A large cloud of red dust filled the air, and then a fine white ash descended to cover the trees and plants. There then followed ear-piercing whistling sounds, three in total, after which three mighty explosions were heard in rapid succession. Immediately after the explosions, the whole forest became a blazing

123

inferno which lasted for several months, depopulating a large area.

The San Calixto Observatory, in operation since 1913, was one of the few places in the region which operated a seismograph (de la Reza 2000a). In fact, one of the best seismological registers of the day, using Galitzin photographic paper, had been put into operation there in 1930 in time for the August event. On this record, three events occurred, the first at 12:04:27, the second at 12:04:51, and the third at 12:04:56 UT. These times correspond to a few minutes after 8 o'clock in the morning local time in the River Curuçá area.

On infrared images taken by the LANDSAT satellite and from aeroplane radar maps, [Ramiro] de la Reza [an astrophysicist at the National Observatory in Rio de Janeiro] identified one major feature to the southeast of the town of Argemiro, near the River Curuçá, which might be an impact signature. The feature corresponds to an astrobleme about 1 km in diameter.

http://cosmictusk.com/mini-tunguska-the-rio-curuca-brazil-1930/

A large majority of meteorites which reach the Earth's surface fall into the oceans unobserved, and only the largest and brightest that traverse the daytime skies are noticed by humans. Of the more than three dozen eyewitness accounts that I reviewed, at least half reported seeing an incoming meteor burst in the air, and most of these reports also included an account of the subsequent noise emanating from that explosion. A few reported experiencing thermal effects, such as feeling a sensation of warmth following the explosion, or finding superheated fragments still in a plastic state that quickly quenched into hard rock.

For over a century, the largest and most famous meteoritic airburst was the one that occurred over Tunguska, Siberia, on June 30, 1908, at 7:17 AM local time. The temperature of the fireball was so intense that some storage huts in the vicinity were severely damaged, and silverware and tin utensils within were deformed by the heat. Eighty million trees were uprooted and blown down and laid radially outward within a 30-40 km radius of "ground zero" (see Figure 67). Some trees on the leeward side of hills were not toppled, but their branches were nonetheless broken off, and bark was stripped off to leave them standing as naked poles. After the airburst, fires ravaged an area of about 10-15 km in radius. The trunks of many trees were not burned through, but only scorched on the surface. This was interpreted by some scientists as evidence of a scorching heat wave passing through the forest instead of a conventional forest fire. Estimates of the energy released by the explosion of the Tunguska bolide range from 3-20 megatons of TNT.[76] At 15 megatons, the energy released by the Tunguska bolide explosion would equal about 1,000 atomic bombs of the size dropped on Hiroshima, Japan.

Credit: Kridec, 1966; Sullivan, 1979
Figure 67 - Footprint of Tunguska Airburst Damage

---

[76] The energy released from one megaton of TNT equals 4.184 petajoules (PJ); the energy from one ton of TNT equals 4.184 gigajoules (GJ).

125

The notoriety of Tunguska was superseded on February 2, 2013, when a bolide exploded over Chelyabinsk, Russia, at 9:09 AM local time.[77] Video imagery of this event, many of which included audio, was captured on almost 700 devices, most of which were dashboard-mounted cameras ("dashcams") in automobiles (see Figure 68). In addition to the dashcams, data about this cosmic intruder was collected from several different sources including: seismic records, infrasonic records from around the world, satellite observations (U.S. Government sensors and meteorological satellites), recovered meteorites (fragments of the bolide), and damage assessments on the ground.

Credit: Borovicka, 2014

Figure 68 - "Dashcam" images of the Chelyabinsk bolide

The energy released by the Chelyabinsk airburst was initially estimated at 500 kilotons (kt) of TNT. The length of the luminous path across the sky was 272 km and its altitude spanned from 95.1 km at ignition to 12.1 km at detonation (see Figure 69). The initial velocity at ignition was about 19 km/sec, and the terminal velocity at detonation was 3.2 km/sec. The total visible transit time was about 16 seconds.

---

[77] The description and images of the Chelyabinsk event were derived from Borovicka, 2014.

Figure 69 - Chelyabinsk Bolide Flight Path (altitude in km)

The initial mass of the asteroid, based upon known energy and speed, was estimated at 12,000 metric tons. The initial size (diameter) of the asteroid, using a meteorite density of 3,300 kg/m^3) was estimated at 19 m. Extreme pressures and thermal stress were placed on the asteroid by the friction of air molecules as it coursed through the atmosphere and eventually caused it to break up. The first fragmentation of the asteroid occurred at an altitude of ~45 km, resulting in a mass loss of about 1%. A large-scale fragmentation, resulting in a mass loss of ~95%, occurred at an altitude of ~35 km. By the time the asteroid reached 29 km, it had been reduced to a collection of 10-20 boulders ranging in size from 1-3 m. The final, explosive fracturing of the asteroid fragments occurred at an altitude of 12.1 km.

Because the explosive detonation of the fragments occurred at a relatively high altitude, most of the ground damage was caused by the shock wave of the combined fragments traveling through the atmosphere at hypersonic speed (see Figure 70). The linear distribution of fragments caused the shock wave to take on a cylindrical form. Secondary, weaker shocks after the arrival of the main shock were also spherical in shape and caused by the explosive fragmentation of the individual remaining pieces.

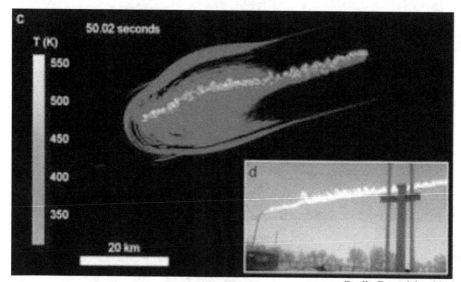

Credit: Borovicka, 2014

Figure 70 - Chelyabinsk Bolide Shock Wave Profile

Over 7,000 buildings were damaged by the Chelyabinsk airburst (see Figure 71). Most of the damage was confined to broken glass in windows, but the roof of a zinc factory also collapsed. Most of the glass damage was caused by atmospheric overpressure from the initial shock wave and subsequent pressure waves from the explosions of the asteroid fragments. Of the more than 5,000 windows examined, about 10% broke due to the initial shock wave; the rest were broken by the subsequent overpressures from the fragment explosions.

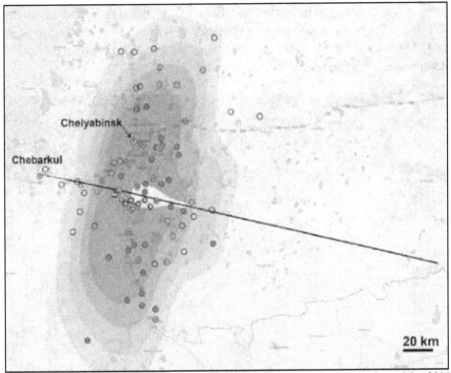

Chelyabinsk

Chebarkul

20 km

Credit: Borovicka, 2014

Figure 71 - Chelyabinsk Glass Damage Map

Over 1,600 people sought medical treatment at hospitals following the airburst, but only 112 were actually hospitalized, including two in serious condition. Most of the injuries were from broken glass, but others reported burns (like a severe sunburn), painful eyes (temporary flash-blindness), temporary deafness, and emotional stress. Fortunately, there were no deaths and no injuries directly related to falling meteorites.

The Chelyabinsk airburst came as a complete surprise for at least two reasons. First, watchers of the night skies were preoccupied with asteroid 2012 DA14, a near-Earth asteroid about 45 m in diameter. This asteroid was expected to pass within 27,700 km (17,200 miles) of Earth on February 15, 2013, which would bring it inside the orbital belt of geosynchronous communications and weather satellites. Asteroid 2012 DA14 approached Earth from below the ecliptic plane and, fortunately, passed safely by Earth without colliding with any of the numerous satellites orbiting Earth in the geo-belt.

The Chelyabinsk asteroid, which is less than half the size of 1999 NC14, approached Earth in its ecliptic plane from the direction of the

Sun, a direction in which the watchers of the night sky had not been looking with their radars and sophisticated instruments (see Figure 72). Also, because it came from the direction of the Sun, they were blind to its approach in the daylight.

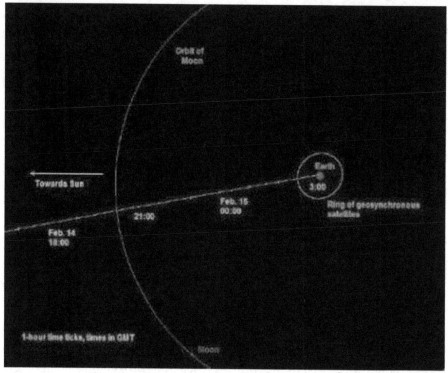

Figure 72 - Chelyabinsk Asteroid Approach

The second reason the watchers of the night sky failed to notice the approach of the Chelyabinsk asteroid is that it's orbit was virtually identical to the known orbit of asteroid 86039, also known as asteroid 1999 NC43 (see Figure 73). This 2-km asteroid intercepts Earth's orbit at two points (roughly mid-October and mid-February of Earth's orbit) but comes within close proximity to Earth only about every seven years. 2013 was not the year of close approach, however, so there was no reason to be watching for asteroid 86039 or anything possibly near it or sharing its orbit.

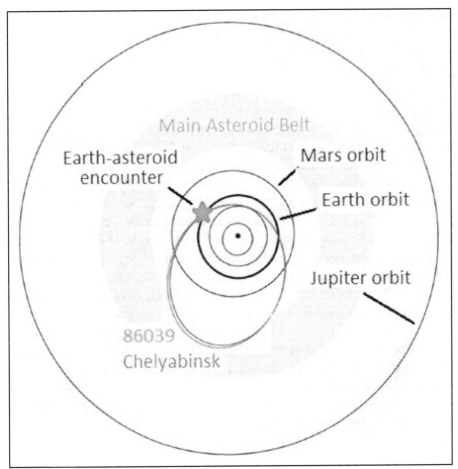

Credit: Borovicka, 2014; redrawn and supplemented by Phil Silvia
Figure 73 - Chelyabinsk Asteroid Orbit

Chelyabinsk asteroid might be a chip knocked off of asteroid 86039 by a collision with another asteroid during one of its transits through the main asteroid belt. When this alleged collision occurred is unknown, however.[78]

It has long been speculated that the destructive nature of a meteoritic airburst closely parallels that of an atomic explosion, but without the nuclear radiation. The Chelyabinsk airburst affirmed this speculation. At approximately 500 kilotons of TNT equivalence, the Chelyabinsk airburst occurred at a high enough altitude that no ground fires were ignited,

---

[78] See http://www.nasa.gov/topics/solarsystem/features/watchtheskies/russian-meteor-nature.html#.VJN9VHs8Gix

and humans experienced nothing worse than the equivalent of a slight-to-moderate sunburn. The Tunguska airburst, however, was closer to 20 megatons of TNT equivalence and ignited, or at least scorched and flattened, over 2,000 km^2 of pine forest.

Concerning airburst phenomena and effects, most of the material damage caused by an atomic bomb or exploding cosmic object is due mainly—directly or indirectly—to the shock front (sudden rise in overpressure) that accompanies the explosion (Gladstone, 1957).[79] The majority of structures will suffer some damage when the atmospheric overpressure exceeds about 3.5%.[80] The shock front overpressure associated with an airburst detonation, however, can exceed ambient pressure by more than 40%. The distance to which this overpressure will extend depends upon the yield of the explosion and the height of detonation. The expansion of the intensely hot gases at extremely high pressures in the ball of fire produced by an atomic explosion causes a blast wave to form in the air that is nominally spherical and expands outwardly at high velocity. In the first moments after the detonation, the variation in pressure with distance from the point of detonation is minimal, and the overpressure at the shock front is nearly twice the overpressure within the fireball, which is of considerable magnitude (see Figure 74).[81] As the shock front travels away from the point of detonation, the overpressure behind the front steadily decreases at a rate of approximately $1/D^2$ where D is the distance from the point of detonation.

---

[79] Despite the date of his publication (1957), Gladstone remains the authority on atomic bomb explosion phenomenology that everyone else quotes. A newer edition of his work has been published, but differs from the original mostly in that the human effects of atomic explosions has been edited out in (what I believe is) the spirit of "political correctness." I chose to reference his original work for the descriptions contained herein.

[80] The atmospheric pressure at mean sea level is defined as 760 mmHg, 29.92 inHg, or 14.696 psi. A 3.5% overpressure is about 26 mmHg or 0.5 psi.

[81] All diagrams in this section were derived from Glasstone, 1957, and have been redrawn by P. Silvia for clarity.

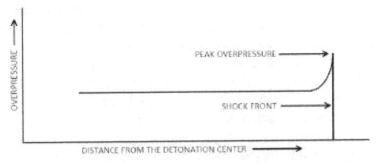

Figure 74 - Overpressure immediately following detonation

From a fixed point on the ground away from ground zero (the point directly below the detonation), no change in pressure is noted immediately following the explosion since it takes time for the shock front to travel the separation distance. When the shock front arrives, the pressure suddenly increases tremendously and is felt by whatever object is encountered by the shock front as an intense, super-heated wind of 300 km per hour (186 miles per hour) emanating from the direction of the blast (see Figure 75). This is the point at which most of the initial damage occurs, first by the shock of the sudden arrival of fast-moving hot air, and second by the flying debris that is carried along by the sustained, hot wind (dynamic pressure) behind the shock front races across the ground.

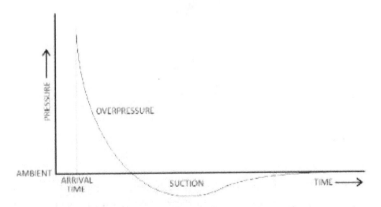

Figure 75 - Variation of pressure over time

Following the arrival of the shock front, the overpressure falls rapidly, as does the force of the wind. A few seconds after the passage of the shock front, the overpressure falls to zero while the now much lighter

133

wind continues to blow in the outward direction. The air that was so rapidly expanded by the heat of the passing shock front begins to cool, and this causes the air pressure to fall below the original ambient level over the next few seconds. The resultant vacuum (underpressure) causes the flow of air to reverse direction. Some additional damage may be caused by flying debris during this phase of underpressure, but it will be significantly less than the damage caused by the passage of the initial shock front.

Although the intensity of the shock front generally decreases at the rate of $1/D^2$ as it moves away from the point of detonation, the dynamics of an airburst actually amplify the overpressure and somewhat negate the decrease caused by distance as the shock front reflects off the surface of the ground (see Figure 76).

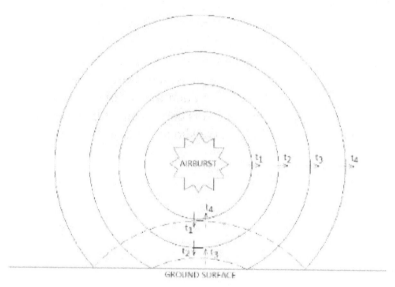

Figure 76 - Shock front reflected from the ground surface

The incident wave of the shock front moving away from the point of detonation travels at a speed governed by the ambient temperature and air pressure. The air temperature and pressure within the sphere of the shock front, however, are significantly higher, and this allows the reflected wave to travel faster. Eventually, the reflected wave overtakes the incident wave, and the two fuse to form a "Mach" or "irregular" reflection or "Mach stem" (see Figure 77).

134

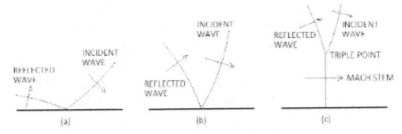

Figure 77 – Formation of the Mach Stem

The point at which the incident and reflected waves fuse to form the Mach stem is called the "triple point." As the reflected wave continues to overtake the incident wave, the triple point rises higher above the ground and the height of the Mach stem increases (see Figure 78). Any object on or above the ground and below the triple point will experience a single shock wave whose intensity equals the sum of both the incident and reflected shock fronts. Any object that is above the triple point will experience the incident and reflected fronts separately.

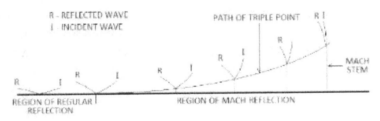

Figure 78 - Movement of the shock front in the Mach stem region

The Mach stem is nearly vertical to the ground, and the transient winds behind it travel parallel to the ground. Thus, in the Mach region, the blast forces on above-ground structures and other objects are directed nearly horizontally. Consequently, vertical surfaces are more heavily loaded and stressed than horizontal surfaces.

The distance from ground zero at which Mach fusion begins to form the Mach stem depends upon the yield of the detonation and the height of the burst above ground. For a typical airburst of 1-megaton yield at a height of 500 meters, the Mach stem begins to form about 2 km from ground zero. As the height of detonation increases, the distance at which Mach fusion commences moves further away. If the detonation occurs at a sufficiently great height, then a Mach stem may not be formed at all.

135

Large, hilly land masses can amplify the intensity of the shock wave if the slope is upward, away from ground zero. Conversely, hilly land masses that slope downward, away from ground zero, can diminish the shock wave intensity. Hills may also "shadow" downrange objects from the thermal energy of the blast, but not the overpressure since the blast waves easily bend, or diffract, around and over obstacles.

Clouds and atmospheric thermal gradients may also serve as reflective surfaces to the expanding sphere of the shock front. For wind velocities that increase by about 8 km per hour for every 500 m increase in altitude, the shock front of the blast will be reflected back to the ground within the first 1,000 m of the atmosphere. As with the formation of the Mach stem, the convergence of multiple reflected waves can form concentrations of blast energy that will greatly exceed the value that would otherwise occur at that distance. A somewhat similar enhancement of pressure (and noise) from large explosions has been reported at greater distances, 100-120 km in winter, and 200-250 km in summer.

As observed with both the Tunguska and Chelyabinsk airburst events, the shape of the shock wave was cylindrical, not spherical. This results in the expanding shock wave and Mach stem on the ground taking on a butterfly-shaped pattern (see Figures 67 and 71, above). The cylindrical shape creates irregular reflection patterns that can produce more points of blast wave concentration than with a spherical shock wave.

Chapter 14. SOIL GENERATION AND MORPHOLOGY

The three major requirements for a civilization to take root and remain in an area are: (1) defensible high ground on which to live and take refuge when enemies approach, (2) reliable sources of water for drinking and crop irrigation, and (3) arable ground on which to grow crops to feed the people and livestock. The extended absence of an occupying civilization following the destruction event begs the question: What was missing that prevented people from returning to the area? The hills were not removed from their places. Springs and surface wadis continued to flow. What, however, had become of the ground?

My initial impression upon seeing Tall el-Hammam and its surroundings for the first time was that the destructive force that removed mudbrick buildings from their foundations may have also damaged the ground sufficiently to prevent the growing of food crops. What might that damage have been, and how long would it have taken to recover from that damage?

Soil scientists use a simple system to describe different soil layers—literally an "ABC" of dirt (see Figure 79). The partially decomposed organic matter found at the ground surface is called the O-horizon. This organic layer, whose thickness varies with vegetation and climate, typically consists of leaves, twigs, and other plant material on top of the mineral soil. The organic horizon may be missing altogether in arid regions with sparse vegetation, whereas in thick tropical jungles the O-horizon holds most soil nutrients.

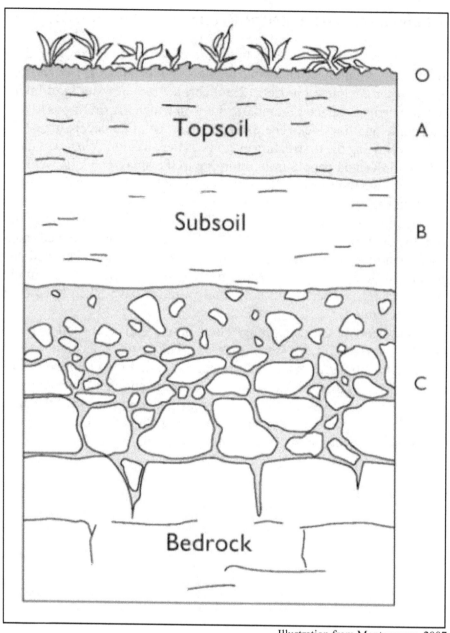

Illustration from Montgomery, 2007
Figure 79 - Soil Layers

Below the organic horizon lies the A-horizon, the nutrient-rich zone of decomposed organic matter mixed with mineral soil. Dark, organic-rich A-horizons at or near the ground surface are what we normally think of as dirt. Topsoil formed by the loose O- and A-horizons erodes easily if exposed to rainfall, runoff, or high winds.

The next horizon down, the B-horizon, is generally thicker than the topsoil, but less fertile due to lower organic content. Often referred to as subsoil, the B-horizon gradually accumulates clays and cations[82] carried down into the soil. The weathered rock below the B-horizon is called the C-horizon.

Concentrated organic matter and nutrients make soils with well-developed A-horizons the most fertile. In topsoil, a favorable balance of water, heat, and soil gases fosters rapid plant growth. Conversely, typical subsoils have excessive accumulations of clay that are hard for plant roots to penetrate, low pH that inhibits crop growth, or cement-like hardpan layers enriched in iron, aluminum, or calcium. Soils that lose their topsoil generally are less productive, as most B-horizons are far less fertile than the topsoil (Montgomery, 2007).

There are five key factors governing soil formation: geology (parent material—rocks), climate, organisms, topography, and time. The geology of a region controls the kind of soil produced when rocks break down, as they eventually must when exposed at the earth's surface. Granite decomposes into sandy soils. Basalt makes clay-rich soils. Limestone just dissolves away, leaving behind rocky landscapes with thin soils and lots of caves. Some rocks weather rapidly to form thick soils; others resist erosion and only slowly build up thin soils. Because the nutrients available to plants depend on the chemical composition of the soil's parent material, understanding soil formation begins with the rocks from which the soil originates (Montgomery, 2007).

How fast soil is produced depends on environmental conditions. Hot temperatures and high rainfall favor chemical weathering and the conversion of rock-forming minerals into clays. Cold climates accelerate the mechanical breakdown of rocks into smaller pieces through expansion and contraction during freeze-thaw cycles. Water slowly percolating down through soils redistributes the new clays, forming primitive mineral soils. Alpine and polar soils tend to have lots of fresh mineral surfaces that can yield new nutrients, whereas tropical soils tend to make poor agricultural soils because they consist of highly weathered clays leached of nutrients. At high latitudes, perpetually frozen ground can support only the low scrub of arctic tundra. Moderate temperatures and

---

[82] Cation—an atom with a positive charge that has been produced by the loss of one or more electrons.

139

rainfall in temperate latitudes support forests that produce organic-rich soils by dropping their leaves to rot on the ground. Drier grassland soils that support a lot of microbial activity receive organic matter both from the recycling of dead roots and leaves and from the manure of grazing animals. Arid environments typically have thin rocky soils with sparse vegetation (thin A-horizon and virtually no O-horizon). Hot temperatures and high rainfall near the equator produce lush rainforests growing on leached-out soils by recycling nutrients inherited from weathering and recycled from decaying vegetation. In this way, global climate zones set the template upon which soils and vegetation communities evolve (Montgomery, 2007; Hillel 1991).

Plant roots, microorganisms, and various bacteria increase weathering rates by releasing CO_2 produced during respiration into soil pore space where combines with water to produce weak carbonic acid. Consequently, rocks buried beneath vegetation-covered soils decay much faster than bare rock exposed at the surface. Once organic matter begins to enrich soils and support the growth of more plants, a self-reinforcing process results in richer soil better suited to grow even more plants (Montgomery, 2007).

The topography, or slope of the land, affects the rate of soil formation. Excessive accumulation of soil in low spots can reduce the rate of new soil formation by burying fresh rock beyond the reach of soil-forming processes. Conversely, erosion strips away soil and exposes bare rock to weathering processes that can either accelerate soil formation or virtually shut it off, depending upon how well local plants can colonize the exposed rock. Given enough time, soil evolves toward a balance between erosion and the rate at which weathering forms new soil (Montgomery, 2007).

For whatever reason, agriculture—the purposeful cultivation of crops—developed independently in Mesopotamia, northern China, and Mesoamerica (Montgomery, 2007; Hillel 1991). The earliest known semi-agricultural people lived on the slopes of the Zagros Mountains between modern Iran and Iraq about 11000-9000 BCE (see Figure 80). By 7500 BCE, herding and cultivation had replaced hunting and gathering as the mainstay of their diet. The earliest evidence for systematic cultivation of grains comes from Abu Hureyra in the headwaters of the Euphrates River in modern Syria. The settlement around Abu Hureyra grew rapidly with the conscription of animal labor around 6000 BCE. Fueled by growing harvests, the population swelled to between 4,000 and 6,000 people.

Population growth kept pace with increasing food production. The increasing population also created a demand for increased food production. It was not long before the impact of topsoil erosion—caused by intensive agriculture and concentrated animal grazing—began to reduce crop yields. Upland erosion and growing population in the Zagros Mountains pushed agricultural communities into the lowlands. Settled villages of up to 25 households spread out across the fertile valleys of the Tigris and Euphrates Rivers. The first farmers had practiced "dryland farming" and relied solely upon rainfall to water their crops. But, once farmers moved into the northern portion of the floodplain between the Tigris and Euphrates rivers, they reaped bigger harvests by digging and maintaining canals to water their fields (Montgomery, 2007; Hillel 1991).

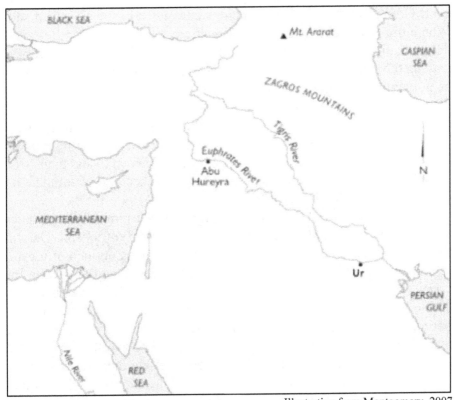

Illustration from Montgomery, 2007

Figure 80 - Map of the Tigris-Euphrates Valley

All of the good, fertile land in Mesopotamia—the land between two rivers—was under cultivation by 4500 BCE. About this same time, the plow was introduced in Sumeria, near the Persian Gulf, and this allowed even greater production from land already under cultivation. Towns began to coalesce into cities as urban centers absorbed surrounding villages. Eventually, eight major urban centers dominated Mesopotamia. When it was founded, the city of Ur was a harbor on the Persian Gulf. Accelerated sedimentation caused by erosion of cultivated soils moved the shoreline of the Gulf southeastward at a rate of almost 30 m per year such that Ur is now about 80 km from the north end of the Gulf (Montgomery, 2007; Hillel 1991).

As urban sprawl removed land from cultivation, the burgeoning population raised demand for increased production from the land that remained under cultivation. Increased irrigation that nourished Mesopotamian fields and helped to boost production carried a hidden risk, however. Groundwater in semiarid regions contains a lot of dissolved salt. Where the water table is close to the surface, as in the river valleys and deltas, capillary action moves groundwater up into the soil to evaporate, leaving the salt behind in the soil. When evaporation rates are high, as in heavily cultivated and irrigated fields, sustained irrigation can generate enough salt to poison crops. Wheat, one of the major Sumerian crops, is quite sensitive to salt in the soil. The earliest temple records from about 3000 BCE report equal amounts of wheat and barley in the region. By 2500 BCE, wheat had dropped from 50% of the harvest to less than 20%. By 2000 BCE, wheat could no longer be grown because of salinization (Montgomery, 2007; Hillel 1991).

The United States Department of Agriculture estimates that it takes five hundred years to produce an inch (about 2.5 cm) of topsoil. Darwin thought English earthworms did much better, making an inch of topsoil in a century or two. When he returned home to England following his voyage around the world, Darwin noticed a layer of fine earth had covered a layer of cinders that he had scattered about his fields before embarking upon his voyage. Upon closer examination, Darwin noticed the resemblance between the castings earthworms bring to the surface and the texture of the soil covering the cinders. Darwin then noticed one of his fields that had last been plowed in 1841, leaving the field covered with exposed rocks that clattered loudly as his young sons ran down the slope. Thirty years later, a horse could gallop across the field without striking a hoof on a rock. By digging a trench in the field, Darwin discovered that the stones were covered by about 2.5 inches of soil, all of which he concluded had been brought to the surface by earthworms. A similar discovery was made in Surrey, where millennium-old Roman ruins and a stash of coins were discovered buried by 6 to 11 inches of soil, implying a formation rate of about one-half to one inch per century (Montgomery, 2007).

While England may have ideal conditions for generating new dirt, soil formation rates vary in different regions of the world according to differences in both geology and climate. Arid and semiarid regions require much longer periods of time to generate soil suitable for plants to grow. Erosion is also another factor that limits the net accumulation of usable soil. Many centuries of accumulated soil can be washed away during a single season of unusually high precipitation (Montgomery, 2007; Hillel, 1991). The explosive force of a meteoritic airburst can accomplish the same scale of soil destruction and removal in a matter of seconds (Gladstone, 1957). If an airburst occurs close to a body of salt water, the vapor plume raised by the detonation can deposit a significant volume of salt brine over a very large area (Gladstone, 1957) and poison the remaining soil for a long time. In semiarid regions, such as the Middle Ghor, the recovery time to regenerate usable soil is pushed to the longer end of the scale because of insufficient precipitation to flush the salt from the soil (Montgomery, 2007; Hillel 1991).

Chapter 15. DISCUSSION OF THE DATA

As stated at the beginning of Chapter 7, there are three assertions in the title of my dissertation that guided the research for my study. These same three assertions also guided the presentation of the data in the previous chapters and will continue in this chapter with the discussion of the data.

The first assertion is in the phrase: "The Middle Bronze Age Civilization"—that a thriving civilization occupied the MG during the MBA, despite the fact that the presence of this civilization in the TMG was unrecognized by archaeologists and historians of the MG until relatively recently.

The second explicit assertion is in the phrase: "Civilization-Ending Destruction"—that something happened to bring that occupying civilization to an end that was not just a minor set-back, but a total annihilation. But, there is also a second part to this assertion, namely, that the consequences of the destruction resulted in a sufficiently long occupational hiatus before people returned to the area.

The third explicit assertion is in the phrase: "Destruction of the Middle Ghor"—that this civilization-ending event involved not just a single site, but an entire region. That destruction was both abrupt and severe, and it took several centuries for the consequences of that destruction to abate to a point where people could reoccupy the area with permanent settlements.

Pre-destruction Occupation of the Middle Ghor

The geological setting of the Middle Ghor is a unique place on earth—inside the Great Rift Valley, the longest and deepest crack in the Earth's crust that is not entirely submerged under one of the great oceans. The micro-climate of the Middle Ghor is also unique to the area, having somewhat warmer temperatures during the winter months and slightly cooler temperatures during the summer months than the high plateaus to the east and west. It was an ideal setting within the Ancient Near East for civilization to take root.

Early archaeologists (most notably Albright, Mallon, and Glueck) totally misread the early occupation history of the TMG. By limiting their exploration to surface surveys, they focused on material remains that were readily visible and accessible on the surface of the ground over which they walked. This led to two failures. The first was failure to recognize and understand the propensity of Iron Age builders to repurpose building materials. This is very obvious at Tall el-Hammam, where soundings excavated in 2013 revealed trenches filled with a homogeneous matrix of dirt where former foundation stones had been removed in Field LS. Excavations in 2014 in Fields LS and UB revealed significant layers of fill dirt that had been used to bury MBA architecture and provide a level surface for new IA construction. Further excavations in Field LS in 2015, which I personally supervised, revealed the removal of over two-thirds of the stones from the then surface MBA foundation walls by the IA builders when they cleared and leveled the surface to reuse those stones for their own construction above the few MBA remnants that they left behind. The stones that the IA builders incorporated into their foundation walls included grinders, socket stones, and other architectural and domestic-use stones used by the previous occupants. Early explorers of the region did not understand that IA builders had buried evidence of earlier occupation, and this led to their second failure. Without the benefit of excavation, they erroneously concluded that the original settlers of the TMG were IA people because theirs was the material evidence that they found on the surface.

Through their excavations, J. Flannigan at Tall Nimrin, K. Prag at Tall Iktanu, T. Papadopoulos at Tall Kafrayn, and S. Collins at Tall el-Hammam have all documented the presence of a thriving and robust civilization continuously occupying the TMG from at least the Chalcolithic period (4600-3600 BCE) through MB2 (1800-1550 BCE). They also concur with the findings of the early explorers that IA people established a notable presence during IA2 (1000-586 BCE) into early IA3 (586-332 BCE), which was subsequently followed by Roman, Byzantine, and Islamic occupation.

These four sites are the largest Bronze Age sites in the TMG, with Tall el-Hammam being the largest by far. Although the numerous smaller sites surrounding them have not been excavated, it is reasonable to assume that they, too, were thriving satellite sites at the same time. The overall occupation picture that emerges from excavations that have been conducted in the TMG since the late 1980s is that of a large and robust MBA city-state community with Tall el-Hammam and possibly Tall Nimrin as centers of hegemony. Substantial agricultural activity would have been required to support the large population. Thus, many smaller towns and villages would have surrounded the larger fortified urban centers. At the height of the growing season, the region would have appeared as a well-watered, lush and fertile valley.

Fresh water sources are significantly less on the Cisjordan Middle Ghor (CMG). Consequently, Jericho is the only known urban center. Kenyon confirmed that it, too, was actively occupied during the MBA (Kenyon, 1957). Thus, thriving centers of human occupation with corresponding agricultural activity and commerce were spread across the entire Middle Ghor during the MBA.

Destruction of the Middle Ghor

Destructive military conquests usually target individual urban centers, but there is neither archaeological evidence nor written record of any destructive military conquest within the Middle Ghor prior to the Late Bronze Age destruction of Jericho.[83] The evidence collected for this study, however, suggests that a destructive event occurred during the Middle Bronze Age that impacted the entire Middle Ghor.

Archaeological Evidence of a Destruction Event

There is archaeological evidence of major earthquake damage to many settlements during the Early, Intermediate, and Middle Bronze Ages. The early permanent settlements of Jericho show evidence of destruction and abandonment at the end of both the Early and Intermediate Bronze Ages with resettlement occurring nearby, but not at the site of the previous settlement.[84] Teleilat Ghassul was abandoned at the end of the Chalcolithic Period and was never resettled. Tall Iktanu shows some evidence of destruction at the end of the Early Bronze Age followed by abandonment. Archaeologists have concluded that the destruction of these sites was the result of earthquakes, and abandonment was the probable result of water sources being either shut off or diverted by the same cause.[85]

---

[83] The primary account of the destruction and conquest of Jericho during the Late Bronze Age is contained Chapter 6 of the book on Joshua in the Bible. See also Kenyon, 1957.

[84] Archaeologists have unearthed the remains of more than 20 successive settlements in Jericho, the first of which dates back 11,000 years (9000 BCE), almost to the very beginning of the Holocene epoch of the Earth's history. See also Kenyon, 1957.

[85] Some researchers have proposed that the destruction observed was the result of military action, but there is no archaeological evidence to support this claim. The most logical cause of the damage is geological, i.e., tectonic movement resulting in earthquakes. See also Prag, 2007.

Tall el-Hammam also shows signs of earthquake damage during the Early and Intermediate Bronze Ages (Collins, *et al*, 2009b). This site did not experience a disruption of its water sources, however, and occupation continued without interruption. Homes were either repaired or rebuilt on the original foundations, and fortification walls were repaired and strengthened (Collins, *et al*, 2010). Heavily damaged cultic buildings were torn down, and new foundations were laid to support the replacement structures (Collins, *et al*, 2014).

Clearly, the effects of the major earthquakes of the Early and Intermediate Bronze Ages were felt across the entire Middle Ghor since evidence of damage has been found from Jericho on the western edge of the CMG to Tall el-Hammam on the eastern edge of the TMG. The Ghor is, after all, a major geological fault. That seismic events would occur at relatively frequent intervals is no surprise. That some of these events would be classified as "major" and cause significant damage is also no surprise. Even today, earthquakes are common in the region. Between 1907 and 1993, nearly 100 earthquakes having magnitudes between 4.0 and 6.9 on the Richter scale were recorded in the Middle Ghor, including the Dead Sea basin (Neev and Emery, 1995).

The destruction event that hit the Middle Ghor at the end of the Middle Bronze Age is considerably different, however, from a number of perspectives. First and foremost is the sheer magnitude of the destruction. During an earthquake, it is very common for buildings made of unreinforced block such as mudbrick to fall down. This is typically the result of lateral movement of the ground. This movement is usually bidirectional, and buildings topple in both directions as the ground moves back and forth.[86] Some of the mudbricks do crumble, but large sections of walls remain relatively intact when the building collapses. Although weaker sections of wall are the more frequent victims of earthquake-induced collapse, thicker and stronger sections usually remain standing.

Few mudbrick wall sections have been found in the MBA destruction layer at Tall el-Hammam. On the south and west sides of the Upper Tall, most of the mudbricks seem to have been reduced to powder or small chunks. Where sections of wall have been found, they mostly have fallen in a single direction, generally to the northeast. Virtually every architectural wall was knocked completely off of its foundation. Mudbrick has

---

[86] Pidwirny, N. (2014) "Earthquake." (http://www.earth.org/view/article/151858).

been found *in situ* only in the remaining bottom few courses of the multi-meter-thick fortification walls or building walls that were sheltered by nearby fortification walls.[87]

Several courses of mudbrick wall have been found *in situ* on the meter-thick foundations of large, monumental buildings on Upper Tall el-Hammam. Such thick walls are the ground floor of what would have been multi-story buildings (palaces, etc.). Consistently, however, the standing mudbrick walls that remain are no higher than the even thicker mudbrick fortification walls that encircle the upper city. The upper stories of these buildings appear to have disappeared completely, leaving no residue of crumbled mudbrick behind.

An ash layer is usually found in association with an earthquake destruction layer. This is to be expected since oil lamps and cooking fires can ignite the fallen organic debris (architectural timbers, wall and floor rugs and mats, etc.). This ash layer is typically characterized by large chunks of charcoal and partially-consumed wood.[88] The ash layer associated with the MBA destruction event at Tall el-Hammam, however, is up to a meter thick in some areas, and the ash is very fine with charcoal bits usually the size of a dried split pea or smaller. This is indicative of a massive conflagration associated with the destruction event, far more extensive and hotter burning than is typically associated with an earthquake event. Whether this extensive burning is the direct result of the destruction or the insufficiency of human intervention (or total lack thereof) cannot be determined, but the extraordinary magnitude and duration of the burning is clear from the depth of ash.[89]

Although extraordinary depths of ash were not reported to have been found at Tall Iktanu or Tall Kafrayn, neither were standing mudbrick walls on building foundations reported from the MBA. Flannigan reported finding nineteen courses of mudbrick on top of six courses of stone foundation that had been exposed by the bulldozer cut that was

---

[87] Some EBA mudbrick has also been found under MBA foundation walls. This mudbrick was purposefully left in place when the MBA foundations were laid and was thus shielded from the MBA destruction event.

[88] Ash layers fitting this description were found on both Upper and Lower Tall el-Hammam during Season Ten (2015).

[89] Some have speculated that the intense burning was triggered by an earthquake-related release of light fractions of hydrocarbons (e.g., methane gas) from the heavier hydrocarbons known to exist in the Dead Sea subsurface (Neev and Emery, 1995). There is no known record, however, of a gas release covering an area as large as the Middle Ghor (~500 km²).

148

made into the north side of the tall for road construction during his survey of Tall Nimrin in 1986 (Flannigan, 1990). No attempt was made to date this architecture, however. Other than remnants of massive fortification walls and a few building that were sheltered by fortification walls, the only MBA architectural features found at any site in the Middle Ghor are foundations made of field stones. When people returned to the area during the Iron Age, they made liberal use of architectural stones left behind by others (rather than haul them in from distant quarries) and brought in fresh alluvial dirt to bury what remained and provide level pads on which to lay their own foundations. New foundations were constructed from recycled stones, and the foundations left behind by previous occupants of the site were rarely left intact or built directly upon by the Iron Age people.

Material Evidence of a High-T Destruction Event

The material evidence included in my study can be divided into three basic categories: pottery, soil/ash, and "melt products." [90] Pottery analysis results for Tall el-Hammam were provided by S. Collins and senior staff members of the Tall el-Hammam Excavation Project. The TeHEP four-step methodology for "reading" excavated ceramics involves a triple-blind analysis protocol. The first step is a staff field reading (SFR) that is performed weekly during the excavation season by members of the senior staff (S. Collins and G. Byers, principal readers). The second step is the Jordanian specialist reading (JSR) that occurs at the close of the season with in-country experts (A. Abu-Shmais and J. Haroun, principal readers). The third step is the outside expert reading (OER) that takes place in the U.S. under the auspices of Trinity Southwest University's College of Archaeology (R. Mullins, principal reader[91]). In the final step (UPR, unified publishable results), the relatively few instances of discrepancies are studied and harmonized in preparation for official publication. This careful analysis of the pottery is used to date the context from which the sherds were recovered. This is a continuing process that

---

[90] Selected pieces of pottery were treated as melt products because of their unusual condition.

[91] Robert Mullins is an associate professor of Biblical Studies at Azusa Pacific University in California. He received his Ph.D. in Archaeology from the Hebrew University of Jerusalem in 2003. He is a leading specialist in ancient Levantine ceramics. He is currently Director of the Tel Abel Beth Maacah Excavation, Israel.

has spanned eleven seasons of excavation at the time of this writing. Pottery analysis for other sites was gathered in summary form from published excavation reports submitted by the respective dig directors, notably K. Prag (1989, 1990, 1991), J. Flannigan (1990, 1992, 1994, 1996), and T. Papadopoulos (2007, 2010, 2011).

Soil/ash and melt product analysis for this study was provided by a team of scientists, professors, and scholars from across the United States (see Appendix D). Collectively, they have invested countless hours examining material from the Younger Dryas Boundary (YDB), a stratum known to be present across North America and parts of South America, Europe, and Asia, often lying directly beneath an organic-rich layer, referred to as a "black mat" (Bunch, *et al*, 2012). The formation of the YDB is believed to have been caused by multiple fragments of an asteroid or comet that impacted Earth along a 12,000+ km swath across five continents about 12,900 years ago and triggered the Younger Dryas cooling event.

The YDB is characterized by numerous markers which include:

- magnetic spherules measuring 10-250 μm (average 30-50 μm), magnetic grains measuring 1-500 μm, iridium and nickel-enriched sediments, and carbon spherules measuring 0.15-2.5 mm (Firestone, *et al*, 2007);
- spherules of melted silica glass and vesicular siliceous glass (called scoria-like objects, or SLOs) (Bunch, *et al*, 2012); and
- an unusually high peak in platinum content, coupled with a very high platinum/iridium (Pt/Ir) ratio (Petaev, *et al*, 2013).

The primary objective of this analysis team was to look for YDB-like markers in the Tall el-Hammam soil and ash samples as indicators of a possible airburst event. They also examined an assortment of melt products (melt rocks and vitrified pottery) collected from Tall el-Hammam and nearby sites.

While many spherulitic particulates have been found, only a few of these have been revealed that possess the dendritic surface morphology and composition typically associated with an impact event and are therefore insufficient to provide any clear indication of an airburst event. While only a few impact associated spherules have been positively identified thus far in the ash and soil samples, they are much smaller than those found in the YDB samples and the surface of many spherule candidates are obscured by a crust-like coating. Thus, many more impact spherules may be present, but it does mean that the analysis of ant sand was unprofitable. A new set of bulk soil and ash samples was collected from the MBA destruction layer on Upper Tall el-Hammam during excavation Season Ten (2015) just prior to submission of this dissertation.[92] More samples need to be collected from other architectural settings.

While no clearly definitive evidence was identified in the microscopy analysis of the sand and ash samples, more than one "smoking gun" was found that point to a probable occurrence of a meteoritic airburst event over the Middle Ghor as the cause of the widespread destruction. The first "smoking gun" is the large melt rock found near Tall Mweis. It is a fused assemblage of three distinct agglomerations of material consisting of silica and quartz sand and other mineral grit. Based upon the large size of the mass (672 grams), it appears that each of the three parts was separately formed in superheated, turbulent air that melted enough of the sand to accrete sand particles into a mass of gritty, taffy-like consistency. After the first two masses impacted together, they fell out of the air and stuck to a rock (foundation stone?) or mudbrick wall. As the exposed surfaces of the combined mass continued to melt, a third mass of material fell into and stuck to the glassy surface. The superheated environment lasted just long enough to melt a thin veneer of glass over the exposed surface of the entire mass and started to flow toward the edges before it abated and the glass quenched. The entire "melt rock" later broke free from its mount and was found in a mixed matrix of dirt, sand, and rocks.

Microscopy analysis of the melt rock revealed melted zircons and quartz grains, both of which have melting temperatures in excess of 2,000° C (Fel'dman, et al, 2006; Gueguen, et al, 2014). Long-term exposure to a heat source sufficient to melt these minerals would have reduced the entire mass of material to molten glass. But, except at the surface, the individual grains of sand were fused and not melted. The heat source, then, had to have been 4-6 times higher than the 2,000° C melt temperature of the zircons and quartz, and the exposure time to this temperature had to be measured in seconds, a few seconds at most.[93] This dynamic of extreme turbulence, temperature, and short duration is found

---

[92] Analytical results from this set of samples are not yet available for inclusion herein.
[93] Correspondence with T. Bunch and A. West.

only in the shock wave of atomic bomb and meteoritic airburst explosions.[94] An atomic bomb is obviously out of consideration in this context; therefore, a meteoritic airburst is a viable candidate as the source of the thermal gradient that produced the melt rock.

A second "smoking gun" is the collection of vitrified pottery sherds that was recovered from Tall el-Hammam. Two distinct modes of vitrification were discovered in these sherds. The first is similar to the large melt rock from Tall Mweis in that only one surface of the sherd had melted to form a glass veneer. The extreme heat required to melt the zirconium and chromium nuggets found just below the glassy surface is indicative of an extremely high temperature gradient. That the thermal gradient penetrated only half way through the 5 mm thickness of the sherd is indicative of an extremely brief exposure time.

The other group of sherds had been transformed into porous, pumice-like texture by the generation of gas pockets within the body of the sherd as certain minerals melted and trapped water vaporized while the sherd was in a partially melted, taffy-like state. The pieces of "pumiceware" are much smaller (< 1 cm^2 vs. ~32 cm^2) than the sherds that experienced single-surface vitrification, which is probably why the thermal gradient was able to penetrate the full 5 mm thickness even though the exposure time was essentially the same.

The temperature required to produce both forms of vitrified pottery had to be in excess of 4,000° C in order to melt and boil the zircon- and chromium-enriched nuggets (Fel'dman, *et al*, 2006; Gueguen, *et al*, 2014; McEwan, *et al*, 2011). To melt just the top 1-2 mm of the larger, 5 mm thick sherd, the exposure time to this temperature had to be measured in milliseconds or the entire sherd would have been either transformed into "pumiceware" or completely melted through its entire thickness.

While the temperatures required to produce the melt rock and pieces of vitrified pottery are not definitive proof of a meteoritic airburst, they are highly suggestive, because these temperatures required a thermal source that was well beyond MBA anthropogenic capability as well as beyond that of typical natural processes such as forest fires. The temperature/time profile required to produce these melt products, however, is well within the capability of a meteoritic airburst.

---

[94] Meteor impacts also produce similar effects, but there is no crater in the Middle Ghor. Therefore, an impact event is not considered to be viable as the source of destruction.

A third "smoking gun" was found in the chemical and mineral analysis of ash and soil samples taken from the MBA destruction layer of Field LS at Tall el-Hammam. The platinum content of the ash layer was found to be 600% higher than the background level of the soil above and below the ash. Elevated levels of platinum are common in asteroid fragments that have fallen to Earth as meteorites. High levels of platinum have also been found in the black mat associated with the YDB (Petaev, et al, 2013). The appearance of a high level of platinum in the MBA destruction ash layer at Tall el-Hammam suggests that a meteoritic airburst may be the ignition source of the massive fires that accompanied the destruction of the city as well as the source of the violent blast that knocked mudbrick buildings off of their foundations.

Chemical analysis of the ash and soil samples also revealed a high concentration of salts, a unique feature not associated with the YDB. The salts were determined to be of the types found in the Dead Sea. The concentration of salts exceeded 60,000 ppm in the ash layer and 45,000 ppm in the soil layers immediately above and below the ash. Salt concentrations above 20,000 ppm are toxic to most plants used for human consumption. Therefore, it can be safely assumed that such a high concentration of salt was not present in the soil prior to the destruction event. Had it been otherwise, it would not have been possible to grow sufficient crops to sustain the then present population.

The obvious source of the salts in the soil is the Dead Sea because of the similarity of chemical composition. Only the method of transport is in question. The highest level of the Dead Sea recorded in the walls of the Great Rift Valley is 185 m below mean sea level (bmsl) and is dated to about 18,000 YrBP. At this high water mark, the concentration of dissolved salts would not even qualify this body of water to be called the "Sick Sea." By the Chalcolithic Period (ca. 6,600 YrBP), the level of the water had fallen to about 300 m bmsl and the concentration of dissolved salts had reached levels that are toxic to living organisms, which earned it the name "Dead Sea." Since then, the level of the Dead Sea has fluctuated between 350 and 450 m bmsl, and its current level is near its historic Byzantine Era (ca. 600 CE) low of ~450 m bmsl (Neev and Emery, 1995).

The highest point on Upper Tall el-Hammam is 124 m bmsl, and the highest point of Lower Tall el-Hammam is 154 m bmsl. No part of Tall el-Hammam has ever been inundated by the waters of the (now) Dead Sea. The alluvial plain surrounding Tall el-Hammam is around 185 m bmsl, but it was last covered with water long before the dissolved minerals and salts in the water reached a level that is toxic to plants and fish.[95] The annual flooding of the Jordan River, upon which ancient agriculture depended, brought an infusion of fresh water and silt over the central Middle Ghor, but it did not raise the level of the Dead Sea high enough to allow its concentrated brine to "backwash" onto the plain. Seasonal flow through the many wadis also helped to flush the salts from the TMG floor with fresh water and deposit new soil from the Jordanian plateau above.

In contrast, a meteoritic airburst provides a probable mechanism for highly salinated water transport from the Dead Sea onto the Jordan plain of the Middle Ghor. If ground zero of a meteoritic airburst was sufficiently close to or over the north end of the Dead Sea, then a substantial plume of water would have been drawn up into the air, atomized, and spread across the landscape as a superheated brine as the blast front spread out. This is precisely the phenomenon that has been observed in the testing of atomic bombs over the South Pacific Ocean following WW II (Gladstone, 1957).

What is clear from the concentration of salts found in the MBA destruction layer at Tall el-Hammam is that a substantial volume of brine was deposited across the entire site. Even now, as excavations are still being conducted at Tall el-Hammam, the momentary increased humidity caused by lower overnight temperatures is sufficient to draw salts from the soil and produce a white crystalline haze that is visible in the morning on soil surfaces exposed the day before. I personally found a 3-5 mm thick crust of salt on the balk (vertical wall surface of an excavation) cut through the MBA destruction layer that was left untouched for three years. Such a high concentration of salts that are chemically identical to the salts of the Dead Sea had to have a mechanism of transport capable of moving the atomized brine over 12 km to reach Tall el-Hammam and

---

[95] It is also probable that the current elevation of the valley floor is significantly higher than it was when the surface of the water was at 185 m bmsl because 18 millennia of alluvial fill has washed down from the hills since the Dead Sea was at its original level.

beyond. A meteoritic airburst near or over the north end of the Dead Sea is a viable candidate for that mechanism.[96]

Geometry of the Destruction Event

The architectural evidence of destruction on Upper Tall el-Hammam is greater on the south and west sides than on the north and east sides. It appears that the steep sides and height of Upper Tall el-Hammam partially sheltered the north side from the force of the blast that removed the mudbrick superstructures of the buildings from their foundations on the south side. On Lower Tall el-Hammam, the few identifiable segments of fallen mudbrick wall that have been identified all fell in a general north-easterly direction. Both of these observations suggest that ground zero of the blast was southwest of Tall el-Hammam.[97]

The temperature of the hypothetical blast shock front could still have been sufficient by the time it reached Tall el-Hammam to melt pottery and produce the vitrified pottery sherds. The intensity was greater at Tall Mweis, 8.5 km SW of Tall el-Hammam, and sufficient to produce the large melt rock. Smaller samples of melt rock were noted at Teleilat Ghassul, 7 km SSW of Tall el-Hammam and 3 km N of Tall Mweis. (Unfortunately, these samples were not collected when the opportunity presented itself before the discovery of similar items at Tall el-Hammam and Tall Mweis.) The relative sizes of the melt products observed at these sites again suggest that Tall Mweis was closer to ground zero of the hypothetical blast than either Teleilat Ghassul or Tall el-Hammam.

The footprint of destruction caused by the 1908 Tunguska airburst, which approached from the ESE, would have completely engulfed the modern city of Rome. Assuming an approach from the SW for an airburst

---

[96] In the abstract to his American Geophysical Union (AGU) paper to be presented in December 2015, Mark Boslough proposed: "The reaction impulse from such an airburst is therefore similar to a much larger non-plume-forming nuclear explosion. Momentum is coupled through the atmosphere to the surface, generating disproportionately large seismic signatures....This result suggests that coupling from an over-water plume-forming airburst could be a more efficient tsunami source mechanism than a collapsing impact cavity or direct air blast because the characteristic time of the plume is closer to that of a long-period wave in deep water."

[97] Collins, *et al*, 2011, contains a vivid description of skeletal remains of two adults and one child who were killed during the MBA destruction event. All three skeletons were encased in a matrix of ash and broken mudbrick. One adult appeared to have fallen near the south wall of the room under a tumble of mudbrick. The other adult and the child appeared to have been thrown violently against the opposite (north) wall of the same room.

over the Middle Ghor, only 60% of the Tunguska footprint is required to cover the entire Middle Ghor (see Figure 81, below).

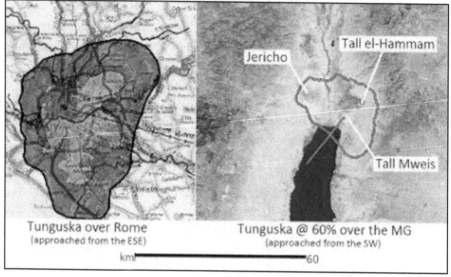

Illustration by P. Silvia

Figure 81 - Tunguska Footprint Comparison

The yield of the Tunguska airburst is estimated at 20 megatons (20 x 10^6 tons) of TNT with a detonation altitude of about 10 km (~6 miles). Because the damage footprint of the Middle Ghor was contained within the valley floor and did not extend beyond the rim of the Great Rift Valley (nominal altitude ~650 m amsl in the vicinity of the MG), I assume the detonation altitude to be approximately 600 m amsl, which is about 1 km above the floor of the Middle Ghor. Thus assumption would account for the apparent minimal impact of the destruction event observed through archaeological excavations on the Jordanian Plateau to the east and the Judean Plateau to the west of the Middle Ghor. A detonation altitude of 1 km does pose some issues, however. First, it would require a relatively step angle of approach to allow the approaching meteoritic body to penetrate deeper into the atmosphere. Second, such a low altitude of detonation is more likely to produce a crater, which does not exist in the Middle Ghor. This suggests that ground zero of the detonation may have been over the north end of the Dead Sea rather than northeast of (but close to) the Dead Sea.

The Tunguska explosion annihilated everything at ground zero and decimated everything else within the blast zone (see Figure 67, above).

A similar result may have occurred within the TMG during the MBA destruction event. It is reasonable to assume, therefore, that the blast pressure at ground zero over the TMG could have been approximately equivalent to the ground zero blast pressure at Tunguska.

The theoretical blast pressure at a distance from a static explosion is proportional to the cube root of the energy yield (Gladstone, 1957). This is expressed mathematically as:

$$\frac{D}{Do} = \sqrt[3]{\frac{W}{Wo}}$$

where: D = the target distance (TMG burst altitude)
Do = the reference distance (Tunguska burst altitude)
W = the target yield (TMG burst yield)
Wo = the reference yield (Tunguska burst yield)

The proportional yield required to achieve the same blast pressure is expressed as:

$$W = Wo \: x \left(\frac{D}{Do}\right)^3$$

Therefore, the theoretical TMG burst yield required to achieve an equivalent Tunguska blast pressure is:

$$W = 20 \: x \: 10^6 \: x \left(\frac{1}{10}\right)^3 = 20 \: x \: 10^6 x \left(\frac{1}{10^3}\right)$$

$$= 20 \: x \: 10^3 = 20 \: kilotons$$

Thus, a 20 kiloton airburst at an altitude of approximately 1 km above the valley floor would have about the same effect at ground zero as the Tunguska airburst.[98]

The forests of Tunguska, however, were a relatively "soft" target for that event, so the focus must now shift to extent of damage versus yield. Gladstone (1957) provided a nomogram and bar chart (see Figure 82) derived from test data to estimate the yield required to produce a specific level of structural damage at a given distance. "Explosion Yield" in the nomogram is scaled from 1 to 20 megatons, and "Distance from Ground

---

[98] As a point of comparison, the atomic bomb dropped on Hiroshima by the U.S. during WWII had an estimated yield of 16 KT and detonated at an altitude of 580 m (Kerr, 1991).

Zero" is scaled from 0.05 miles (0.08 km) to 50 miles (80.5 km). The "Construction Line" scale refers to the accompanying bar chart of "Structural Type." The bar chart provides a damage scale for various structural types for surface bursts ("SB") and airbursts ("AB").[99]

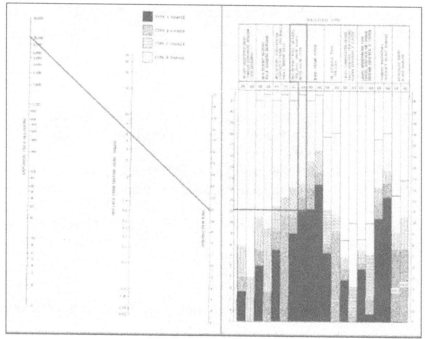

Credit: Gladstone, 1957

Figure 82 - Damage-Distance-Yield Relationships

To estimate the airburst yield required to destroy ("Type A" damage) the buildings on Tall el-Hammam, I am assuming that a MBA mudbrick structure is approximately equivalent to an unreinforced 1950s cinderblock structure from which Gladstone derived his test data. I also assume that ground zero for the MBA airburst was located near the northeast corner of the Dead Sea, about 10 km (6 miles) from Tall el-Hammam. From Gladstone's bar chart, the "Construction Line" value for total destruction of unreinforced brick structures is 11. Drawing a straight line on the nomogram from 11 on the "Construction Line" scale through 6 miles on the "Distance from Ground Zero" scale results in an estimated

---

[99] Gladstone stated that the AB assessment was for a "typical airburst" but did not specify an altitude. Both atomic bombs dropped over Japan during WWII were detonated at an altitude < 1 km, which I am assuming is Gladstone's "typical" altitude, and is equivalent to my estimated MBA airburst altitude.

"Explosion Yield" of 10,000 kilotons (10 megatons), or about one-half the maximum estimated yield of the Tunguska airburst.

Gladstone's test data was derived from atomic bomb testing conducted on a flat, wide-open test range. The Middle Ghor, although flat on the valley floor, is surrounded by steep rises up to the Jordanian plateau on the east and the Judean highlands on the west. The steep valley walls would have contained and amplified the effects of the blast within the MG, thus requiring a lower yield to achieve the same effect (Gladstone, 1957).

It should also be noted that the estimated altitude of most cosmic airburst events is > 10 km versus the 1 km that I have proposed here. The apparent lack of damage above and beyond the rim of the Middle Ghor has driven me to this hypothesis. While the comparisons of the Tunguska event and static atomic bomb tests to the Middle Ghor event may not be totally analogous, they at least provide a reasonable gauge for estimating the magnitude of the energy release required to produce the extent of destruction observed at Tall el-Hammam and the other three excavated sites in the Middle Ghor.

Occupational Hiatus in the Middle Ghor

A common pattern was discerned in the excavation reports from across the entire Middle Ghor that reveals a remarkable consistency for the sites that were occupied during the MBA. The evidence affirms that the introduction of new pottery forms of all types ceased near the end of the MBA (1950-1550 BCE), in MB2 (1800-1550) BCE. The dating is based on the evolution of the piriform juglet and the absence of the cylindrical form that appeared in late MB2 and LB1 (1550-1400 BCE) elsewhere in the Levant.[100] In the CMG, the appearance of new pottery forms resumed during the LBA after a lapse of 250-300 years. In the TMG, the appearance of new pottery forms resumed during the late IA1b (1100-1000 BCE) in the area around Tall Nimrin and during early IA2a (1000-900 BCE) in the area around Tall el-Hammam.[101] Thus, the lapse in the appearance of new pottery forms in the TMG lasted 600-700 years. This

---

[100] Personal communication with S. Collins, who has been conducting a study of the evolution of the piriform juglet for future publication.

[101] Distinct LBA pottery forms were found during Season Ten (2015) on Upper Tall el-Hammam in a very isolated context. No LBA pottery forms have been found anywhere else on Tell el-Hammam that would indicate a large-scale LBA resettlement of the site.

gap in the appearance of pottery forms suggests a corresponding absence of permanent human settlements during that period which is affirmed by the general lack of LBA architecture and artifacts.

Resettlement of the Middle Ghor

Archaeological evidence indicates that permanent human settlement of the CMG resumed at Jericho around the beginning of LB2 (1400-1200 BCE) but disappeared again within a few decades.[102] Permanent settlement of the TMG resumed in the north at Tall Nimrin for a brief period of a few decades around the middle of IA1b (1100-1000 BCE) and in the central part at and near Tall el-Hammam around the beginning of IA2a (1000-900 BCE). It seems to be a reasonable hypothesis that the primary reason for the occupational hiatus was damage to the soil rendering it agriculturally unviable. If correct, it follows that recovery of the soil is what allowed permanent human resettlement to return. A brief (2-3 decades) return of human occupation in the CMG resumed after a lapse of about 300 years, but human occupation of the TMG did not resume until after a lapse of 600-700 years. This suggests that the damage to the soil caused by the destruction event was greater in the TMG than it was in the CMG.[103]

Since there were no permanent settlements in the TMG during the occupational hiatus, there is no archaeological evidence available to use in determining the recovery process that occurred. There is only one source of literary information available—the Bible—that at least gives a few clues from its narrative accounts.

In Genesis 50, following the death of his father Jacob, Joseph is granted permission by Pharaoh to take his father's body back to the land of Canaan for interment with his wife and ancestors. Because of Joseph's position as vizier in Egypt, a huge entourage of Egyptian dignitaries, chariots and horsemen accompanied Joseph and his relatives. The relevant part to this discussion is in verses 10 & 11:

> When they reached the threshing floor of Atad, near the
> Jordan, they lamented loudly and bitterly; and there Jo-

---

[102] The end of this brief reoccupation was the result of conquest (Kenyon, 1957), possibly by the Israelites after their Exodus from Egypt (Joshua 6:1-20).

[103] If Tall el-Hammam is actually the ancient city of Sodom, as Collins asserts (Collins and Scott, 2013), then "bad ju-ju" (religious superstition) may also have contributed to the extended occupational hiatus.

seph observed a seven-day period of mourning for his father. When the Canaanites who lived there saw the mourning at the threshing floor of Atad, they said, "The Egyptians are holding a solemn ceremony of mourning." That is why that place near the Jordan is called Abel Mizraim.

The field of Machpelah, which contained the cave in which Jacob was to be entombed, is near Mamre, south of Jerusalem (Genesis 23). The text, however, says that Joseph and his entourage stopped at a place called "the threshing floor of Atad" to mourn Jacob for seven days. The phrase translated "threshing floor" also has the connotation of a "high place."[104] Lower Tall el-Hammam is indeed a high (about 30 m above the surrounding plain), flat place that would have made a suitable threshing floor were it not for two things: first, there is no archeological evidence of grain crops being grown and threshed in the area during the time of Joseph (late MB2, ca. 1550 BCE); and second, the word *atad* tells us what was actually growing there at that time.

The King James translators of these verses treated *atad* as a proper name and transliterated it as "Atad" instead of translating it. Most English versions since then have followed this same tradition and kept the name "Atad." In three other appearances elsewhere in the Old Testament, the word *atad* is translated "bramble" (Judges 9:14, 15) or "thorns" (Psalms 58:9). There is no logical reason, therefore, why the King James translators treated *atad* as a proper name in the Joseph narrative instead of translating it as "bramble" or "thorns" to be consistent with its other three appearances. Thus, "the threshing floor of Atad" would be better translated "the high place of thorns (or thorn bushes)." This would have been an apt description of Lower Tall el-Hammam[105] at that time since thorn bushes are among the first of the colonizing species to return to a site following a cataclysmic destruction because of their ability to grow in poor quality, highly saline soil (Lowe, *et al*, 2000).

---

[104] Theological Wordbook of the Old Testament (TWOT). Chicago: Moody Press, 1980.

[105] Upper Tall el-Hammam is too narrow at the top to accommodate the entourage that accompanied Joseph and the adult Israelite men. Also, its steeply sloped sides were probably as barren of foliage then as they are today.

In both verses 9 and 11, this location is described as being "beyond (or across the ford of[106]) the Jordan." This description always uses Jerusalem as the point of reference; hence, the high place of thorns would have to be on the east side of the Jordan above the flood plain. Once again, Tall el-Hammam is the most likely candidate for the high place of thorn bushes.

The "Canaanites who lived there" were probably nomadic shepherds since there is no archaeological evidence of permanent occupation near or on Tall el-Hammam from the time of Joseph. Because of the hubbub and large number of people garbed in Egyptian attire, the Canaanites called the place *Abel Mizraim*—"Mourning Place (*Abel*) of the Egyptians (*Mizraim*)."

In Numbers 22, the Israelites dropped down from the desert east of Moab into the "plains of Moab" approximately 38 years after their exodus from Egypt and "camped along the Jordan across from Jericho" (verse 1). The "plains of Moab" is widely recognized as another name for the TMG (Aharoni, 1979). This would have occurred around 1400 BCE, the beginning of LB2a (Collins, 2005). Later in Numbers, this location is more explicitly identified as *Shittim* (25:1) or *Abel Shittim* (33:49). *Abel*, as noted above, means "mourning" or "place of mourning." *Shittim* means "acacia"[107] and is descriptive of the thorny trees that overgrew the Middle Ghor and especially the TMG during its long period of abandonment following the destruction of the cities of the plain. Acacia trees are a larger version of thorn bushes. They, too, are an invasive species that is able to colonize poor, desert-like soil, but have a slightly lower salt tolerance than thorn bushes (Carson). Their appearance suggests a reduction in the salinity of the soil as it slowly recovered.

Shittim is the place from which Joshua sent the spies to check out Jericho before the Israelites moved on to capture it (Joshua 2:1). According to Joshua 3:15, the Jordan was at flood stage when the Israelites set out toward Jericho, which means that the central Middle Ghor was flooded 3-5 km on either side of the main river channel. Hence, the Israelite encampment had to have been on high ground, which is a primary reason why most biblical scholars identify Lower Tall el-Hammam is as *Abel Shittim* (Aharoni, 1979).

In 1 Kings 9 and 2 Chronicles 8, King Solomon (end of IA1 to the beginning of IA2, ca. 1000 BCE) is credited with building store cities

---

[106] TWOT.
[107] International Standard Bible Encyclopedia (ISBE). Grand Rapids: Eerdmans, 1986.

throughout the land of Israel. Most of these cities are not identified by name in the text. These store cities were used for storing grain and other commodities in stone and plaster-lined "silos" that were dug deep into the hillsides. Small fortresses were built along with the silos to house a troop of soldiers and administrators to protect the grain and manage its distribution. More than two dozen such silos and a fortress have been identified on Upper Tall el-Hammam. All of these silos were cut deep into the MBA ramparts and EBA structures below and were dated to the time of Solomon and later kings of Judah and Israel by the large quantity of mostly IA2 pottery sherds that has been recovered from the silos.

These narratives represent a chronological record of the processes that led to the recovery of the land in the TMG that led, in turn, to its resettlement. It took at least a century or more for salt to be purged from the soil enough to allow thorn bushes to get established. It took another 2-3 centuries for acacia trees to become established. It took more than six centuries for the decaying leaves of these invasive species and other weeds to decompose and form a new A-horizon sufficiently deep to support the resumption of crop growth. Barley has a higher toleration for salt than wheat (USDA, 2011), and is most likely the first of the cereal grains to be reintroduced into the area following the destruction event. By IA2a (1000-900 BCE), the ground had sufficiently recovered to produce surplus crops that needed to be stored and protected. The numerous storage silos at Tall el-Hammam attest to the recovery of the land and its ability once again to produce cereal grain crops.

Resettlement of the north TMG occurred during IA1b (1100-1000 BCE) (Flannigan, et al, 1994a) and spread south as the raising of grain crops resumed. Small villages were established to house the field workers. Small fortresses were built on nearly every hill to house soldiers assigned to protect the stored grains.

Tall el-Hammam's IA settlement included both monumental and domestic architecture and was comparable in size with contemporary Jerusalem, however, the seat of government and control was located in Jerusalem. In contrast to their former city-state status, the IA settlements of the TMG during this period were primarily military outposts, and that is what the early explorers of the TMG found during their surface surveys in the late 19th and early 20th centuries CE.

Chapter 16. ANALYSIS AND APPLICATION

My study delved into three distinct disciplines that previously had not been combined to tell the story of an area that has long been ignored and misunderstood, the Transjordan Middle Ghor. The field of *archaeology* provided insight into the occupation history of the Middle Ghor; the field of *meteoritics* (supplemented by the field of *atomic weapons*) provided insight into a possible, if not probable, cause behind the end of the occupying civilization; and *soil science* provided insight into factors that contributed to the extended occupational hiatus and the conditions that changed to allow civilization to return.

Correlation of Site Occupational History

Survey reports published by early explorers of the Middle Ghor (Albright, Mallon, Glueck) unanimously declared the occupying civilization in the area east of the Jordan River to be of Iron Age origin. This assertion has now been demonstrated to be entirely incorrect. Excavations since the 1980s at Tall Nimrin (Flannigan), Tall Iktanu (Prag), Tall Kafrayn (Papadopoulis), and Tall el-Hammam (Collins) have all discovered and documented the presence of permanent human occupation that goes back as far as the late Neolithic to early Chalcolithic Periods (ca. 4600 BCE). Tall el-Hammam itself shows clear evidence of continuous occupation lasting almost 3,000 years until it abruptly ended around 1650 BCE.

Tall el-Hammam could not have grown to the size encompassed by either the EBA fortification walls or the later MBA fortification walls without an extensive array of smaller towns and villages surrounding it to house the field workers and provide the other services needed to support such a large urban center. Tall Kafrayn, Tall Rama, Tall Tahuneh, and Tall Iktanu are the four largest sites in the immediate vicinity of Tall el-Hammam, but there were undoubtedly many lesser satellite communities nearby as well. Although smaller, the same could be said of Tall Nimrin with its satellites of Tall Mustah and Tall Bleibel.

Based on this new evidence, the history of the Middle Ghor, and especially the TMG, must be rewritten to correct the false picture that has prevailed for the past 150 years. Maps for the region covering the Middle Bronze Age must also be updated with the new information that has been brought to light through the last 30 years of excavation and analysis.

The occupational history of the TMG during the LBA remains essentially unchanged as a result of this study—there were no permanent settlements in the TMG during this period. Nomadic shepherds undoubtedly traversed the area. Caravans continued to use the trade routes that crisscrossed the area and were monitored by isolated outposts such as the one found on Tall el-Hammam in 2015. Virtually nothing was left behind that could be interpreted as evidence of settlement. At best, the nomadic people only established temporary encampments and then moved on as the seasons changed.

Early explorers did at least get it right that the Iron Age saw a proliferation of small forts and outposts established on nearly every hilltop across the TMG. This was not, however, the initial settlement of the TMG, as they claimed, but the reoccupation of the TMG following 6-7 centuries of no permanent settlements.

Correlation of Samples Analysis

Although the early explorers of the TMG failed to identify any substantive material evidence of Bronze Age occupation through their surface surveys, excavations conducted since the 1980s have recovered significant quantities of pottery evidence bearing witness to a robust civilization living in the region during the Early, Intermediate, and Middle Bronze Ages. It is also noteworthy that these same excavations all report an absence of pottery from the Late Bronze Age. In all cases, except for one isolated building on Upper Tall el-Hammam, pottery forms from the MBA were immediately followed by new pottery forms from the Iron Age.

Vitrified pottery samples have been identified thus far only at Tall el-Hammam, and all of these samples were collected from MBA contexts. Although no vitrified pottery was reported from Tall Iktanu, Tall Kafrayn, or Tall Nimrin, it is not known if vitrified sherds were actually found but discarded without comment because of their unusual condition.

The melt rock that was analyzed for this study was a "surface find" at Tall Mweis, so it has no verifiable archaeological context. The unique nature of this item, however, made it easy to identify as a melt product. Analysis of this sample led to the realization that it is not anthropogenic (i.e., not man-made); therefore, I and the researchers who examined it concluded that it is most likely the result of a meteoritic airburst. The extreme thermal and turbulent conditions required to produce the melt rock are akin to the conditions required to produce the vitrified pottery samples found at Tall el-Hammam. This led to the further conclusion that the melt rock and the vitrified pottery were most likely created by the same thermal event, and that the melt rock, because of its size, was closer to the point of origin for the thermal energy, i.e., "ground zero" of the airburst event.

The ant sand collected from Tall el-Hammam and seven other sites in the vicinity did not yield any conclusive evidence of a widespread thermal event. Examination of the ant sand at New Mexico Tech did not rule out such an event, either. All that can be said with surety is that no melt products were found in the samples that I collected. While many spherulitic particles were recovered from the samples from the MBA ash layer at Tall el-Hammam, only a few of these appeared to be melt products (magnetic micro-spherules), too few to be considered conclusive evidence of a meteoritic airburst event.

However, chemical analysis of ash samples from the MBA ash layer at Tall el-Hammam did identify a significant spike in platinum content. This is one of the signature markers (proxies) of a meteoritic airburst or impact. Chemical analysis performed by collaborating researchers also identified a salt abundance enhancement in the ash layer. The salts seen in the ash and adjacent soil were identified as being of the same types as contained in the Dead Sea. This analysis also quantified the concentration of salts as being well above the threshold of toxicity to most vegetation and especially food crops. The sudden appearance of a salt-enhanced layer argues for some causal event to produce the enrichment.

Although the material samples collected from across the TMG for this study were not sufficient to demonstrate conclusively that a meteoritic airburst is responsible for the termination of MBA human occupation, the condition of the MBA structures examined through excavation at the top four (by size) sites in the TMG strongly suggests that the mudbrick superstructures were knocked off their foundations. It is plausible that this destruction was the result of passage of an intense shock front generated by a meteoritic impact and/or airburst. A meteoritic surface impact is implausible as there is no apparent crater within the Middle Ghor. Therefore, a meteoritic airburst is the most likely cause of the destruction.

Correlation of Occupational History with Samples

The timeline of human occupation of the Middle Ghor is defined primarily through the progression of pottery forms and secondarily through the evolution of architecture. The changes in architecture appear in steps that are generally spaced in multi-decade increments. The primary determining factor for such changes is the occurrence of earthquakes in the region that force restoration and rebuilding of existing structures. A secondary factor is population growth which results in the remodeling and expansion of existing domestic structures and construction of new ones. The dating of these architectural changes is accomplished through the examination of the pottery forms contained in the sherds that are embedded in stone foundation walls or incorporated in the dirt matrix that is used to create pads for new construction, floors inside buildings, or pathways and roads around the buildings.

Occupied sites that have experienced the typical scenario of building-destruction-abandonment (BDA)[108] tend to show clear horizons in the progression of pottery forms because each incident of destruction is usually followed by at least a brief period of abandonment before people reoccupy the site. This is the pattern observed at Jericho in the CMG. Of interest in the TMG is that none of the four excavated sites show a BDA profile. There are no clear horizons of change in the pottery forms, but a continuous, overlapping progression of changes from the Chalcolithic Period or Early Bronze Age right into the Middle Bronze Age. This is indicative of continuous occupation and presents a picture of the current generation continuing to use grandma's pottery until it finally breaks and is discarded, to be recycled as filler in the next construction project.

The big break in this pattern occurred in MB2 where the introduction of new pottery forms suddenly stopped. The magnitude of the destruction that occurred at this time resulted in a period of abandonment that lasted about three centuries in the CMG and six-to-seven centuries in the TMG. The appearance of new pottery forms during the second half of the Late Bronze Age in the CMG, albeit briefly, and in the Iron Age in the TMG creates clear horizons in the timeline of human occupation of the Middle Ghor.

---

[108] The "BDA" sequence was initially coined by S. Collins and is defined in *The Tall al-Hammam Excavations, Volume One*, authored by S. Collins, C. Kobs, and M. Luddeni (Eisenbrauns, 2015).

Chapter 17. CONCLUSIONS

The strength of any hypothesized conclusion is derived from the degree to which its components convincingly answer the underlying questions. In addressing the conclusions of this study and analysis, I shall follow the general outline presented in Chapter 4.

Regarding the initial research question: *What happened to bring the Middle Ghor Civilization to such an abrupt and enduring termination?* The archaeological evidence points to a violent and complete destruction of habitat that resulted in the termination of human occupation in the Middle Ghor. Mudbrick superstructures were sheared off their foundations and reduced to rubble and dust. All organic material was reduced to ash except for a few timbers trapped under piles of debris being entirely converted to charcoal. The ground was also poisoned with a distinct layer of highly concentrated salts from the Dead Sea which prevented the resumption of agriculture following the destructive event.

Regarding the subsidiary questions on the extent of the destruction event:

(a) *Based upon the archaeological evidence, how widespread was the destruction of this event?* Archaeological evidence reveals that the destruction caused by this event affected at least the entire Middle Ghor, from Jericho on the western edge of the CMG to Tall el-Hammam on the eastern edge of the TMG.

(b) *What regional differences can be observed within the Middle Ghor?* The damage caused by the destructive event on the CMG resulted in an occupational hiatus that lasted 250-300 years. The damage caused by the destructive event on the TMG resulted in an occupational hiatus that lasted 600-700 years.[109]

Regarding the subsidiary questions on the scope of the destruction event:

(a) *Based upon the archaeological evidence, what specific damage was incurred during the destruction event?* The destructive event resulted in a patterned demolition of standing mudbrick buildings and severe surface damage to the much thicker exposed mudbrick fortification walls and towers. The exposed mudbrick walls of buildings appear to have been sheared off by a destructive blast, whereas some sections of

---

[109] The reoccupation of the CMG occurred ca. 1400 BCE and lasted for only a few decades. Therefore, the occupational hiatus of the CMG was essentially the same as the TMG.

building walls that were protected by the thicker defensive walls survived. Considering the magnitude of the destructive blast required to produce the observed architectural damage, it is possible that all organic material and crops standing in the fields would likely have been incinerated (a large-scale *loss on ignition*). It is also likely that the topsoil which had supported the large urban population would have been blown away by the force of such a destructive event or incinerated by its heat. The remaining subsoil was drenched with a brine of Dead Sea salts that poisoned the ground for agricultural use.

(b) *How does the damage compare site-to-site?* Based upon an admittedly preliminary investigation, the pattern of complete demolition of MBA architecture appears to be consistent across all three of the MBA sites that have been excavated (occupation of Tall Iktanu ended at the end of the EBA), as documented by Flannigan, Papadopoulos, and Collins, who all reported Iron Age architecture being built over the partial remains of MBA foundations.

Regarding the subsidiary questions on the nature of the destruction event:

(a) *What possible phenomena might have caused such widespread destruction?* Three possibilities have been proposed since the late 19th Century. First, that a major earthquake was responsible for the architectural damage and resulting fires. Second, that an earthquake-induced release of hydrocarbon gases ignited, adding its concussive force to the damage caused by the earthquake, and further exacerbating the resulting fires. The third and most recent proposal is that a comet or asteroid fragment exploded in the air over the Middle Ghor with the force of an atomic bomb that leveled the buildings and incinerated all organic material.

(b) *What physical evidence is available to characterize the destruction?* Evidence characterizing the destruction has been found in many forms: 1) Widespread, total demolition of MBA buildings. 2) Meter-thick layers of fine ash which are not typical of earthquake damage. 3) A specific layer of highly concentrated salts in the soil that produce a crystalline haze on freshly exposed surfaces and whose toxicity rendered the area uninhabitable for want of tillable land. 4) An extended occupational hiatus revealed through the material remains (pottery) found in the area.

(c) *What is the probable cause of the destruction?* Earthquakes may cause predictable patterns of destruction but typically not total demolition of structures and their compositional materials. The fireballs produced by clouds of hydrocarbon gases released from geological fissures are generally confined to small areas, and open-air ignition produces a thermal front but not a concussive shock front. Such fireballs also lack the thermal intensity required to produce the melt products found at Tall el-Hammam and Tall Mweis. The event that caused the destruction observed across the Middle Ghor had to deliver both a major concussive shock front and extreme thermal energy. An exploding volcano may produce such effects, but there are no such sources in or near the Middle Ghor. The only other possibility is the impact or airburst explosion of a meteor. There is no impact crater in the Ghor, so the most probable cause of the destruction seems to be a meteoritic airburst. Material evidence gathered to date is not conclusive, but is consistent with an airburst. More research is needed to confirm this hypothesis.

Regarding the subsidiary questions on the consequences of the destruction event:

(a) *What might have caused the extended occupational hiatus?* People living outside of the Middle Ghor who witnessed the destructive event may have allowed religious superstition to prevent their moving into the area to replace the exterminated population. Beyond "bad juju" there is an organic reason why people could not return and establish permanent settlements in the aftermath of the destruction, and that reason is the destruction of the topsoil and poisoning of the subsoil with Dead Sea salts which removed the land from agricultural production.

(b) *What changed that made people able and willing to return?* After centuries of rain and annual inundation of the land by the flooding of the Jordan River, enough of the salts were flushed from the soil to allow invasive species of plants to become established. These plants provided the compost needed to rebuild enough topsoil to eventually allow the resumption of human agricultural activity and the growing of grain crops. The elapsed time required for this change to take place also allowed "bad juju" and superstition to fade from the corporate memory of the people.

The materials analysis conducted for this study was based on a limited collection of samples. While strong indicators of a possible meteoritic airburst were found, it is not possible at this time to say definitively that the cause of the Middle Bronze Age civilization-ending destruction of the Middle Ghor was a meteoritic airburst. All of the melt products that were examined were collected during the first nine seasons of the Tall el-Hammam Excavation Project. Season Ten was concluded just prior to submission of my dissertation as a draft document. The additional materials collected during Season Ten still need to be examined, and the results of that examination may either affirm the claims made herein or demonstrate the need for further materials collection and analysis.

The strength of the argument would also be improved by expanding the search for the MBA destruction horizon throughout the Middle Ghor and further down the eastern shore of the Dead Sea as far as Bab edh-Dhra. This wider search for evidence of the destruction event could be a study in itself.

SUMMARY

The existence of a robust civilization occupying the entire Transjordan Middle Ghor during the Middle Bronze Age was unknown to early explorers of the region but has now been confirmed through a correlation of excavation reports published since the 1980s. Also revealed through these reports and ongoing excavations at Tall el-Hammam is the sudden, violent, and total destruction of the entire region during the second half of the Middle Bronze Age that brought an end to the occupying civilization and left the region uninhabited for nearly seven centuries.

The discovery and analysis of vitrified pottery sherds at Tall el-Hammam and a large melt rock consisting of glass-coated, fused sand suggest that the destruction event was explosive in nature and included an intense shock front and an extreme thermal profile. Based on the mineral content of these melt products, the temperature and exposure time profile required to create them was far beyond the technological capabilities of Middle Bronze Age people. The only known natural source of such an energy profile is a meteoritic airburst. A literal fire came down from the sky and destroyed the cities of the plain, burned every living plant as well as the ground itself, and poisoned what soil remained with a toxic brine of salts from the Dead Sea. This hypothesized event is also broadly consistent with the recorded and presumably eyewitness account in Genesis 19:22-28. (See Appendix E for a full discussion of the biblical text.)

Without usable ground on which to grow crops, it was impossible for people to reestablish permanent settlements in the region immediately or soon after the destruction event. It took over six centuries for the ground to recover sufficiently to support agricultural use again. The city-state that existed in the region during the Middle Bronze Age was replaced during the Iron Age with a series of forts and military outposts of the relatively new kingdom of Israel. An entirely new civilization reoccupied the Transjordan Middle Ghor.

BIBLIOGRAPHY

Abed, A. M., K. Moumani, and K. Ibrahim
> 2001   *Geology of Jordan: Field Guidebook*. Amman: Jordanian
> Geologists Association.

Aharoni, Y.
> 1979   *The Land of the Bible: A Historical Geography*. Philadel-
> phia: Westminster Press.

Albright, W. F.
> 1924   "The Archaeological Results of an Expedition to Moab and
> the Dead Sea." BASOR 14: 2-12.
> 1925   "The Jordan Valley in the Bronze Age." AASOR VI (1924-
> 25): 13-74.
> 1961   *The Archaeology of Palestine*. London: Penguin.

Bandel, K. and E. Salameh
> 2013   *Geologic Development of Jordan: Evolution of its Rocks
> and Life*. Amman: University of Jordan Press.

Belknap, G. E., Captain U.S.N.
> 1883   "A Singular Meteoric Phenomenon." *Science*, Vol. 1, No. 1
> (Feb 9, 1883), 4-6.

Ben-Tor, A. (ed.)
> 1992   *The Archaeology of Ancient Israel*. New Haven: Yale Uni-
> versity Press.

Bestmann, M., G. Pennacchion, G. Frank, M. Goken, and H. de Wall
> 2011   "Pseudotachylyte in muscovite-bearing quartzite: Cosmic
> friction-induced melting and plastic deformation of quartz."
> *Journal of Structural Geology*, Vol. 33, 169-186.

Borovicka, J.
> 2014   "The Chelyabinsk event – what we know one year later."
> Astronomical Institute of the Academy of Sciences of the
> Czech Republic. (PowerPoint presentation
> (http://www.oosa.unvienna.org/pdf/pres/stsc2014/tech-
> 04E.pdf) from the Fifty-first session of the Scientific and
> Technical Subcommittee of the United Nations Committee

on the Peaceful Uses of Outer Space (UNCOPUOS) (10-21
February 2014) (http://www.oosa.unvi-
enna.org/oosa/en/COPUOS/stsc/2014/index.html).

Brazo, M. W. and S. A. Austin

1985   "The Tunguska Explosion of 1908." Geoscience Research
Institute.
[Online: Institute for Creation Research,
http://www.icr.org/research/index/researchp_sa_r05/]

Bunch, T. E., R. E. Hermes, A. M. T. Moore, D. J. Kennett, J. C.
Weaver, J. H. Wittke, P. S. DeCarli, J. L. Bischoff, G. C.
Hillman, G. A. Howard, D. R. Kimbel, G. Kleteschka, C.
P. Lipo, S. Sakai, Z. Revay, A. West, R. B. Firestone, and
J. P. Kennett

2012   "Very high-temperature impact melt products as evidence
for cosmic airbursts and impacts 12,900 years ago." *PNAS*,
Early Edition.

Burdon, D. J.

1959   *Handbook of the Geology of Jordan.* Amman: Government
of the Hashemite Kingdom of Jordan.

Carson, C.

----   "The Disadvantages of an Acacia Plant." Retrieved from SF
Gate Home Guides (*http://homeguides.sfgate.com/disad-
vantages-acacia-plant-69909.html*).

Collins, G. S., H. J. Melosh, and R. A. Marcus

2005   "Earth Impact Effects Program: A Web-based computer
program for calculating the regional environmental consequences of a meteoroid impact on Earth." *Meteoritical &
Planetary Science*, 40, No. 6, 817-840.

Collins, S.

2002a "Explorations on the Eastern Jordan Disk." *Biblical Research Bulletin*, Volume II, Number 18. Albuquerque: Trinity Southwest University, 2002

2002b "Terms of Destruction for the Cities of the Plain." *Biblical
Research Bulletin*, Volume II, Number 16. Albuquerque:
Trinity Southwest University, 2002

2005   *Let My People Go!*. Albuquerque: Trinity Southwest University Press.

2014   *The Kikkar Dialogues*. Albuquerque: Trinity Southwest University Press.

2014   *The Search for Sodom and Gomorrah*. Albuquerque: Trinity Southwest University Press.

Collins, S., K. Hamdan, A. Abu-Dayyeh, A. Abu-Shmais, G. A. Byers, K. Hamdan, H. Aljarrah, J. Haroun, M. C. Luddeni, and S. McAllister

2008   "The Tall el-Hammam Excavation Project, Season Activity Report, Season Three: 2008 Excavation, Exploration, and Survey." Filed with the Department of Antiquities of Jordan, 13 February 2009.

Collins, S., K. Hamdan, G. A. Byers, J. Haroun, H. Aljarrah, M. C. Luddeni, S. McAllister, Q. Dasouqi, A. Abu-Shmais, and D. Graves

2009a   "The Tall el-Hammam Excavation Project, Season Activity Report, Season Four: 2009 Excavation, Exploration, and Survey." Filed with the Department of Antiquities of Jordan, 27 February 2009.

2009b   "Tall al-Ḥammām: Preliminary Report on Four Seasons of Excavation (2006-2009)" *ADAJ* 53: 385-414.

Collins, S., K. Hamdan, G. A. Byers, J. Haroun, H. Aljarrah, M. C. Luddeni, S. McAllister, A. Abu-Shmais, and Q. Dasouqi

2010   "The Tall el-Hammam Excavation Project, Season Activity Report, Season Five: 2010 Excavation, Exploration, and Survey." Filed with the Department of Antiquities of Jordan, 31 January 2010.

Collins, S. and H. Aljarrah

2011   "Tall el-Hammam Season Six, 2011: Excavation, Survey, Interpretations and Insights." Filed with the Department of Antiquities of Jordan, 2011.

Collins, S., Y. Eylayyan, G. Byers, and C. Kobs

    2012   "Tall el-Hammam Season Seven, 2012: Excavation, Survey, Interpretations and Insights." Filed with the Department of Antiquities of Jordan, 2012.

Collins, S., K. Tarawneh, G. Byers, and C. Kobs

    2013   "Tall el-Hammam Season Eight, 2013: Excavation, Survey, Interpretations and Insights." Filed with the Department of Antiquities of Jordan, 2013.

Collins, S. and L. C. Scott

    2013   *Discovering the City of Sodom*. New York: Howard Books, Division of Simon & Schuster.

Collins, S., G. A. Byers, C. M. Kobs, and P. Silvia

    2014 "Tall el-Hammam Season Nine, 2014: Excavation, Survey, Interpretations and Insights." Filed with the Department of Antiquities of Jordan, 2014.

Collins, S., C. M. Kobs, and M. C. Luddeni

    2015   *The Tall al-Hammam Excavations: Volume One—An Introduction to Tall al-Hammam with Seven Seasons (2005-2011) of Ceramics and Eight Seasons (2005-2012) of Artifacts*. Winona Lake, IN: Eisenbrauns.

Conder, C. R.

    1889   *The Survey of Eastern Palestine, Vol I*. The Committee of the Palestine Exploration Fund.

Cooke, B.

    2013   "The Meteor Environment and Spacecraft". Meteoroid Environment Office of NASA. (PowerPoint presentation, received through personal correspondence with Mark Mulholland).

Cooper, E.

    2010   *10,000 Years of Pottery*. Philadelphia: University of Pennsylvania Press.

Dornemann, R.

    1983   *The Archaeology of the Transjordan in the Bronze Ages*. Milwaukee: Milwaukee Public Museum.

Eby, N., Hermes, R., Charnley, N., Smoliga, J. A.

    2010   "Trinitite—the atomic rock". *Geology Today*, Vol. 26, No. 5. Blackwell Publishing, The Geologists' Association & The Geological Society of London.

Fel'dman, V. I., L. V. Sazonova, and E. A. Kozlov

    2006   "Some Peculiarities of Impact Melts (Natural and Experimantal Data)". *Lunar and Planetary Science*, Vol. 37. Department of Petrology, Moscow State University, Moscow, Russia.

Firestone, R., A. West, and S. Warwick-Smith

    2006   *The Cycle of Cosmic Catastrophies: How a Stone-Age Comet Changed the Course of World Culture*. Rochester, VT: Bear & Company.

Firestone, R. B., A. West, J. P. Kennett, L. Becker, T. E. Bunch, Z. S. Revay, P. H. Schultz, T. Belgya, D. J. Kennett, J. M. Erlandson, O. J. Dickenson, A. C. Goodyear, R. S. Harris, G. A. Howard,J. B. Kloosterman, P. Lechler, P. A. Mayewski, J. Montgomery, R. Poreda, T. Darrah, S. S. Que Hee, A. R. Smith. A. Stich, W. Topping, J. H. Wittke, and W. S. Wolbach

    2007   "Evidence for an extraterrestrial impact 12,900 years ago that contributed to the megafaunal extinctions and the Younger Dryas cooling." *PNAS*, Vol. 104, No. 41, 16016-16021.

Flanagan, J. W., D. W. McCreery, and K. N. Yassine

    1990   "First Preliminary report of the 1989 Tell Nimrin Project." *ADAJ* 34: 131-152.

    1992   "Preliminary Report of the 1990 Excavation at Tell Nimrin." *ADAJ* 36: 89-112.

    1994a "Tell Nimrin: Preliminary Report of the 1993 Season." *ADAJ* 36: 205-227.

    1994b "Tell Nimrin: The Byzantine Gold Hoard from the 1993 Season." *ADAJ* 38: 245-265.

    1996   "Tall Nimrin: Preliminary Report on the 1995 Excavation and Geological Survey." *ADAJ* 40: 271-292.

Glasstone, S., ed.

    1957   *The Effects of Nuclear Weapons*. U.S. Atomic Energy Commission.

Glueck, N.

    1934   "Explorations in Eastern Palestine, I." *AASOR* XIV for 1933-1934: 1-114.

    1935   "Explorations in Eastern Palestine, II." *AASOR* XV for 1934-1935.

    1939   "Explorations in Eastern Palestine, III." *AASOR* XVIII-XIX for 1937-1939.

    1951   "Explorations in Eastern Palestine, IV." *AASOR* XXV-XXVII for 1945-1949.

Goethe-Institut

    1991   *Geology of Jordan*. Amman: Al Kutba.

Goodman, J.

    2011   *The Comets of God: New Scientific Evidence for God*. Tucson: Archaeological Research Books.

Graves, D. and S. Stripling

    2011   "Re-examination of the Location for the Ancient City of Livias." *LEVANT* 43, No. 2: 178-200.

Grimal, N.

    1992   *A History of Ancient Egypt*. Oxford: Blackwell Publishers.

Guequen, E., J. Hartenstein, and C. Fricke-Begemann

    2014   "Raw material challenges in refractory application." Berliner Kongerenz Mineralische Neben produkte und Abfälle (conference paper).

Harland, J. P.

    1961   "Sodom and Gomorrah." Pp. 41-75 in *The Biblical Archaeologist Reader*, ed. G. Ernest Wright and David Noel Freedman. Garden City, NY: Doubleday.

Heinrichs, Till, Elias Salameh, and Hani Khouri

    2013   "The Waqf as Suwwan crater, Eastern Desert of Jordan: aspects of the deep structure of an oblique impact from reflection seismic and gravity data." *International Journal of Earth Science* (2014) 103: 233-252. Published online by *Springer* (2013).

Hennessy, J. B.

    1969   "Preliminary Report on the First Season of Excavations at Teleilat Ghassul." *LEVANT* 1: 1-24.

    1982   "Teleilat Ghassul: Its Place in the Archaeology of Jordan." *SHAJ* I: 55-58.

Hillel, D. J.

    1991   *Out of the Earth: Civilization and the Life of the Soil.* New York: Free Press, A Division of Macmillan.

Hoerth, A. J.

    1998   *Archaeology & the Old Testament.* Grand Rapids: Baker Academic.

Hoerth, A. J., G. L. Mattingly, and E. M. Yamauchi

    1994   *Peoples of the Old Testament World.* Grand Rapids: Baker Books.

Ibrahim, M. A., J. A. Sauer, and K. Yassine

    1976   "The East Jordan Valley Survey, 1975." *BASOR* 222: 41-66.

Jones, A., G. Matthew, and L. Cormody

    2013   "Carbonite Ments and Carbonatites." *Reviews in Minerology & Geochemistry* 75: 289-322.

Jones, H. L., trans.

    1930   *Strabo: Geography.* Book XVI, 2:42. Cambridge, MA: Harvard University Press.

Josephus F.

    --   *Antiquities.* I. 11:1-4

    --   *Jewish Wars.* IV. 476-485

Kafafi, Z. A.

    2007   "Late Bronze Age Settlement Patterns North of the az-Zarqā' River." *SHAJ* IX: 389-396.

Kenyon, K. M.

    1957   Digging Up Jericho: The Results of the Jericho Excavations 1952-1956. New York: Frederick A. Praeger.

Kerr, E. B.

    1991   *Flames over Tokyo: the US Army Air Forces' Incendiary Campaign against Japan 1944-1945.* New York: Donald I Fine.

Kridec, E. L.

    1966   *Giant Meteorites.* Oxford: Pergamon Press.

Khouri, R. G.

    1988   *The Antiquities of the Jordan Rift Valley.* Amman: Al Kutba.

Kirkwood, D.

    1879   "On Meteoric Fireballs Seen in the United States during the Year Ending March 31, 1879." *Proceedings of the American Philosophical Society*, Vol. 18, No. 103 (Jan-Jun, 1879), 239-247.

Kitchen, K. A.

    2003   *On the Reliability of the Old Testament.* Grand Rapids: W. B. Eerdmans.

Kuntz, G. F., J. Torrey, and E. H. Barbour

    1890   "The Winnebago County (Iowa) Meteorites." *Science*, Vol. 15, No 380 (May 16, 1890), 305-305.

Le Losq, C., D. Neuville, R. Moretti, and J. Roux

    2012   "Determination of water content in silicate glasses using Raman spectrometry: Implications for the study of explosive vulcanism." *American Minerology*, 97: 779-790.

Lewis, J. S.

    1999   *Comet and Asteroid Impact Hazards on a Populated Earth: Computer Modeling.* San Diego: Academic Press.

Longo, G.

2007   "Chapter 18. The Tunguska Event" in *Comet/Asteroid Impacts and Human Society: An Interdisciplinary Approach.* P. Bobrowsky and H. Rickman, eds. Springer.

Lowe, S., M. Browne, S. Boudjelas, M. De Poorter

2000   *100 of the World's Worst Invasive Species: A selection from the Global Invasive Species Database.* The Invasive Species Specialist Group (ISSG), a specialist group of the World Conservation Union (IUCN). Retrieved from http://www.k-state.edu/withlab/consbiol/IUCN_invaders.pdf .

Mallon, A. (J. R. Duncan, trans.)

1932   "The Five Cities of the Plain." *PEFQ* JAN: 52-56.

1933   *"Duex Fortresses au Pied des Monts de Moab."* *Biblica* 14, 400-407.

Mallon, A., R. Koeppel, and R. Neuville

1934   *Teleilat Ghassul I.* Rome.

Mangerud, J., Andersen, S. T., Burglund, B. E., and Donner, J. J.

1974   "Quaternary Stratigraphy of Norden, a Proposal for Terminology and Classification." *Boreas*: 109–128. http://onlinelibrary.wiley.com/doi/10.1111/j.1502-3885.1974.tb00669.x/abstract

Mazar, A.

1990   *Archaeology of the Land of the Bible: 10000 to 586 BCE.* New York: Doubleday.

McEwan, N., T. Courtne, R. A. Parry, and P. Knupfer

2011   "Chromite—A cost-effective refractory raw material for refractories in various metallurgical applications." (conference paper)..

Merrill, S.

1883   *East of the Jordan: A Record of Travel and observation in the Countries of Moab, Gilead, and Bashan During the Years 1875-1877.* New York: Charles Scribners & Sons.

Miller, A. M.

    1919   "The Cumberland Falls Meteorite." *Science, New Series,* Vol. 49, No. 1275 (Jun. 6, 1919), 541-542.

Montgomery, D. R.

    2007   *Dirt: The Erosion of Civilizations.* Berkeley: University of California Press.

Neev, D. and K. O. Emery

    1995   *The Destruction of Sodom, Gomorrah, and Jericho: Geological, Climatological, and Archaeological Background.* Oxford: Oxford University Press.

Nininger, H. H.

    1928   "Another Kansas Meteorite." *Transactions of the Kansas Academy of Science,* Vol. 31 (Feb 17, 1922 - Apr 14, 1928), 91-94.

Oldfather, C. H., trans.

    1979   *Diodorus of Sicily,* 12 Vols., LOEB Classical Library. Cambridge, MA: Harvard University Press.

Papadopoulos, T. J.

    2007   "The Hellenic Archaeology Project of the University of Ioannina in Jordan: A Preliminary Synthesis of the Excavation Results at Ghawr aṣ-Ṣāfī and Tall al-Kafrayn (2000-2004)." *SHAJ* IX: 175-191.

    2010   "Preliminary Report of the Seasons 2005-2008 of Excavations by the University of Ioannina at Tall al-Kafrayn in the Jordan Valley." *ADAJ* 54: 283-310.

    2011   "Tall al-Kafrayn: The University of Ioannina Hellenistic-Jordan Expedition, Preliminary Report on the Ninth Excavation Season (2009)." *ADAJ* 55: 131-146.

Patten, D. W.

    1966   *The Biblical Flood and the Ice Epoch: A Study in Scientific History.* Seattle: Pacific Meridian. Online: html - www.creationism.org/patten/ and a pdf version at http://www.scribd.com/doc/15782993/The-Biblical-Flood-and-the-Ice-Epoch - scribd.com.

1988   *Catastrophism and the Old Testament: The Mars-Earth Conflicts.* Seattle: Pacific Meridian.

Patten, D. W., R. R. Hatch, and L. C. Steinhauer

1973   *The Long Day of Joshua and Six Other Catastrophes: A Unified Theory of Catastrophism.* Seattle: Pacific Meridian.

Patten, D. W. and S. R. Windsor

1997   *The Mars-Earth Wars (Ending in 701 B.C.E.).* Seattle: Pacific Meridian. [Online: http://www.creationism.org/patten/PattenMarsEarthWars/PattenMEW01.htm]

Petaev, M. I., S. Huang, S. V. Jacobsen, and A. Zindler

2013   "Large Pt anomaly in the Greenland ice core points to a cataclysm at the onset of Younger Dryas." *PNAS*, Vol. 110, No. 32, 12917-12920.

Politis, K. D.

1989   "Excavations at Dier `Ain `Abata 1988." *ADAJ* 33: 227-233.

Prag, K.

1988   "Kilns of the Intermediate Early Bronze-Middle Bronze Age at Tell Iktanu Preliminary Report, 1987 Season." *ADAJ* 32: 59-72.

1989   "Preliminary Report on the Excavations at Tell Iktanu, Jordan, 1987." *LEVANT* 21: 33-45.

1990   "Preliminary Report on the Excavations at Tell Iktanu, Jordan, 1989." *ADAJ* 34: 119-130.

1991   "Preliminary Report on the Excavations at Tell Iktanu and Tell al-Hammam, Jordan, 1990." *LEVANT* 23: 55-66.

2007   "Water Strategies in the Iktānū Region of Jordan." *SHAJ* IX: 405-412.

Prag, K. and H. Barnes

1996   "Three Fortresses on the Wadi Kafrain, Jordan." *LEVANT* 28: 41-61.

Rainey, A. F. and R. S. Notley

2006   *The Sacred Bridge.* Jerusalem: Carta.

Rast, W. E. and R. T. Schuab

   1980   "Preliminary Report of the 1979 Expedition to the Dead Sea Plain, Jordan." *BASOR* 240: 21-61.

Redford, D. B.

   1992   *Egypt, Canaan, and Israel in Ancient Times.* Princeton, NJ: Princeton University Press.

Renfrew, C. and P. Bahn

   2011   *Archaeology Essentials: Theories, Methods and Practice.* London: Thames & Hudson.

Salameh, E., H. Khouri, and W. U. Reimold

   2013   "Drilling the Waqf as Suwwan impact structure." (unpublished; accepted for publication by *Springer*, 2013).

Salameh, E., H. Khouri, W. U. Reimold, and W. Schneider

   2008   "The first large meteorite impact structure discovered in the Middle East: Jabel Waqf as Suwwan, Jordan."*Meteoritics and Planetary Science* 43. No. 10, 1681-1690 (2008).

Schaub, R. T. and W. E. Rast

   1989   "Bâb edh-Dhrâᶜ : Excavations in the Cemetery Directed by Paul W. Lapp (1965-67)." Winona Lake, IN: Eisenbrauns.

Short, T.

   1741   "An Account of Several Meteors." *Philosophical Transactions (1683-1775)*, Vol. 41 (1739-1741), pp. 625-630).

Sullivan, W.

   1979   *Black holes, the edge of space, the end of time.* Garden City, NY: Doubleday.

USDA

   2011   "Crop Toleration and Yield Potential of Selected Crops as Influenced by Irrigation Water Salinization (ECw) or Soil Salinity (ECe)." USDA Natural Resources Conservation Services, Technical Note, June 2011.

Von Haidinger, W. R.
1868 "On the Phenomena of Light, Heat, and Sound accompanying the fall of Meteorites." *Proceedings of the Royal Society of London*, Vol. 17 (1868-1869), 155-160.

Waheeb, M.
1997 "Report on the Excavations at Wādī al-Kufrayn Southern Ghors (al-Aghwār)." *ADAJ* 41: 463-468.

Whiston, W., trans.
1974 *The Works of Flavius Josephus*, trans, in Four Volumes (Grand Rapids: Baker Book House).

Winkless III, N. and I. Browning
1975 *Climate and the Affairs of Men*. Burlington, VT: Fraser Publishing

Winthrop, J.
1762 "An Account of a Meteor Seen in New England." *Philosophical Transactions (1761-1762)*, Vol. 52 (1761-1762), 6-16.

Yassine, K. (ed.)
2011 *Tell Nimrin: An Archaeological Exploration*. Jordan: University of Jordan.

Yassine, K., J. A. Sauer, and M. M. Ibrahim
1988 "The East Jordan Valley Survey, 1976 (Part Two)", in *Archaeology of Jordan: Essays and Reports*, Khair Yassine, ed. Amman: University of Jordan, 189-207.

APPENDIX A—Spelling Conventions

The journal articles and reports referenced for this writing yielded a significant variety of spellings for units of measurement, compass point directions, archaeological sites, locations, and geographical features. For the sake of simplicity, a standardized spelling convention has been used herein, regardless of how the original appeared in quoted text. The following table shows the spelling convention used and the variant spellings that were observed in the source documents. Note, also, that American (versus British) word spelling has been used throughout.

Convention Used	Variants Observed
`Abata	`Abaṭa
Abu Qarf	Abū Qarf
Azeimeh	el-`Aẓeimeh
Bab edh-Dhra	Bâb edh-Dhrâᶜ
Bleibel	Bleibîl, el-Bleibil
el-Ghanam	el-Ghanâm
ghor	ghōr, ghwar
el-Hammam	al-Ḥammâm, el-Ḥammâm (aka el-Ḥammeh es-Samrī)
el-Heri	el-Herī
Hesbon	Ḥesbân
Iktanu	Iktanû, Iktānū
Kafrayn	el-Kefrein, Kefrein
Lisan	Lisān
Metabi	Meṭâbiᶜ
Mustah	Muṣtâh, el-Muṣtâh
Mweis	Mwais
Nimrin	Nimrîn, Nimrīn
Plain of Moab	`Arbôth Mô´áb
qattara	qaṭṭara
Qetein	Qeṭein
Rama	er-Râmeh
as-Safi	aṣ-Ṣāfī
Sha`ib	Sha`îb
Shittim	haš-Šiṭṭîm
Shunah	Shūnah, Shûneh

Tahuneh	eṭ-Ṭāḥûn, Tahuna
Tall	Tel, Tell
Teleilat Ghassul	Tulayāt al-Ghassūl; Tuleilat Ghassul
wadi	wâdī
Zarqa	Zarqā, Zerqā
zor	zōr

APPENDIX B—Acronyms and Abbreviations

The acronyms and abbreviations used throughout this writing are presented in the following tables. For consistency, the directional and unit of measurement abbreviations are also used within quotations regardless of the form that appeared in the original text.

Acronym	Definition
AASOR	*The Annual of the American Schools of Oriental Research*
ACOR	American Center for Oriental Research
ADAJ	*Annual of the Department of Antiquities of Jordan*
AJ	*Astrophysical Journal*
amsl	above mean sea level[2]
ANE	Ancient Near East[-ern]
ASOR	American Schools of Oriental Research
AU	Astronomical Unit
BASOR	*Bulletin of the American Schools of Oriental Research*
bmsl	below mean sea level[2]
BRB	*Biblical Research Bulletin*, a publication of TSU
CMG	Cisjordan Middle Ghor
CP	Chalcolithic Period[1]
DoA	Department of Antiquities
EB; EBA	Early Bronze; Early Bronze Age[1]
EDS	Energy Dispersive Spectroscopy
FIB	Focused Ion Beam
IA	Iron Age[1]
IB; IBA	Intermediate Bronze; Intermediate Bronze Age
IJES	*International Journal of Earth Sciences*
JADIS	Jordan Antiquities Database and Information System
LB; LBA	Late Bronze; Late Bronze Age[1]

LEVANT	*Journal of the British School of Archaeology in Jerusalem and the British Institute at Amman for Archaeology and History*
MB; MBA	Middle Bronze; Middle Bronze Age[1]
MG	Middle Ghor
MPS	*Meteoritics & Planetary Science*
PEFQ	*Palestine Exploration Fund Quarterly*
PN	Pottery Neolithic [Period] [1]
PNAS	*Proceedings of the National Academy of Sciences*
ppm	parts per million
PPN	Pre-Pottery Neolithic [Period][1]
SEM	Scanning Electron Microscopy
SHAJ	*Studies in the History and Archaeology of Jordan*
SLOs	scoria-like objects
TeHEP	Tall el-Hammam Excavation Project
TMG	Transjordan Middle Ghor
TSU	Trinity Southwest University
TWOT	*Theological Wordbook of the Old Testament*
WDS	Wavelength Dispersive Spectroscopy
YDB	Younger Dryas Boundary
YrBP	Years before present

Notes:
1. See Appendix C for dates of the archaeological periods and sub-periods.
2. Referenced to the level of the Mediterranean Sea.

Directions and Units of Measure	Definition
C	Centigrade
cm	centimeter(s)
g	gram
kg	kilogram(s)
km	kilometer(s)
m	meter(s)
mm	millimeter(s)
μm	micrometer(s); micron(s)
N-E-S-W	North-East-South-West

APPENDIX C—Archaeological Periods

The names and dates for archaeological periods used throughout this writing have been adopted from the Tall el-Hammam Excavation Project led by Dr. Steven Collins and are defined in the following table (Collins, Kobs, and Luddeni, 2015).

Age	Abbr	Date Range	Comments
Pre-Pottery Neo-lithic	PPN	8500-6600 BCE	
Pottery Neolithic	PN	6600-4600 BCE	
Chalcolithic Period	CP	4600-3600 BCE	
Early Bronze Age	EBA	3600-2500 BCE	
Early Bronze 1	EB1	3600-3000 BCE	
Early Bronze 2	EB2	3000-2700 BCE	
Early Bronze 3	EB3	2700-2500 BCE	
Intermediate Bronze Age	IBA	2500-1950 BCE	
Intermediate Bronze 1	IB1	2500-2200 BCE	Old EB4
Intermediate Bronze 2	IB2	2200-1950 BCE	Old MB1
Middle Bronze Age	MBA	1950-1550 BCE	
Middle Bronze 1	MB1	1950-1800 BCE	Old MB2a
Middle Bronze 2	MB2	1800-1550 BCE	Old MB2b/c; time of Patriarchs
Late Bronze Age	LBA	1550-1200 BCE	
Late Bronze 1	LB1	1550-1400 BCE	Israel in Egypt
Late Bronze 2a	LB2a	1400-1300 BCE	Exodus; sojourn; con-quest
Late Bronze 2b	LB2b	1300-1200 BCE	Period of the Judges
Iron Age	IA	1200-332 BCE	
Iron Age 1a	IA1a	1200-1100 BCE	Period of the Judges
Iron Age 1b	IA1b	1100-1000 BCE	United Kingdom (Saul/David/-Solomon)

Iron Age 2a	IA2a	1000-900 BCE	Divided Kingdom
Iron Age 2b	IA2b	900-700 BCE	(Judah/Israel)
Iron Age 2c	IA2c	700-586 BCE	Judah only
Iron Age 3	IA3	586-332 BCE	Persian Period
Hellenistic Period	332-63 BCE		
Roman Period	63 BCE - 135 CE		

APPENDIX D—Materials Analysis Team

As word of this project spread, several academicians, scientists, and researchers offered assistance with analyzing the materials that I brought back from Jordan for my research effort. Thus, the materials analysis team was formed. Listed below in alphabetical order by last name are the members of whom I am aware. There may be others, but their identities remain hidden behind the closed doors of the labs in which the materials examinations were performed. Without their enthusiastic assistance, the completion of this study would have been long delayed. Thank you, all, for the generous contribution of your time and resources to this effort.

Dr. A. Victor Adedeji is an associate professor of physics in the department of Natural Sciences at Elizabeth City State University in Elizabeth City, NC. His research interest includes: development of contact metallization scheme for Wide Band Gap (WBG) semiconductor microelectronics useful for special applications in harsh environment; and the deposition and characterization of thin films for "smart window" applications. Drs. Adedeji and LeCompte have been the main team at ECSU studying the mineralogy of impact products (spherules, SLOs, etc.) using a scanning electron microscope (SEM) with an energy-dispersive spectrometry (EDS) x-ray analyzer to examine the material.

Dr. Ted Bunch is Adjunct Professor of Geology at Northern Arizona University, founder and CEO of Space Sciences Consulting Services, and a Research Associate of the Royal Ontario Museum, Canada. He received BA and MS degrees in Geology from Miami University, Oxford Ohio and a PhD from the University of Pittsburgh, Department of Space Sciences. He has over fifty years of research experience in the investigations of experimentally and naturally shocked materials, asteroid impact craters, Apollo lunar samples, terrestrial mineralogy and petrology, meteorites, origin of life, space habitation osteoporosis, exobiology, space environment hazards, and interplanetary dust particles. During his last ten years at NASA, he developed hypervelocity impact techniques for the Ames Vertical Gun hypervelocity impact experiments on meteorites, carbonaceous materials, and aerogel cosmic particle collectors, in addition to characterizations of Martian, Lunar, and asteroid meteorites. He has published over 270 peer reviewed research reports, book chapters, and articles including 88 papers on meteorites, 78 on impact craters and hypervelocity collisions, 44 on the Moon and Mars, and 38 on terrestrial geology. He received NASA's Exceptional Scientific Achievement Medal, among other awards, in recognition of his accomplishments. He is a Fellow in two professional scientific societies, a National Science Foundation Research Fellow, and a NASA-Ames Associate Fellow. An asteroid 7326 TedBunch (1981 UK22) was named after him in recognition of his space science achievements. He retired from NASA Ames Research Center in 2001 as Chief of Exobiology. Dr. Bunch is one of the principal researchers and analysts for the YDB investigation and co-author of several YDB articles.

Dr. T. David Burleigh is Professor of Materials & Metallurgical Engineering at New Mexico Tech, Socorro, NM. Dr. Burleigh began his career in metallurgy by taking a minor in Physical Metallurgy at Colorado School of Mines, and completing both his M.Sc. (1980) and Ph.D. (1985) in Metallurgy at M.I.T. He completed his post-doctoral studies at the Max-Planck Institute in Berlin, Germany, working for Prof. Dr. Heinz Gerischer in photoelectrochemistry. From Germany he moved back to the USA to work at Alcoa Technical Center as a senior and staff engineer in the Alloy Technology Group. He next had various positions at the University of Pittsburgh Materials Science and Engineering Department before moving to New Mexico Tech in 2001. He is a registered Professional Engineer in Metallurgy in New Mexico. At New Mexico Tech he teaches undergraduate engineering classes, and graduate level classes in "Electrochemical Techniques & Processes," "Corrosion Phenomena" and "Failure Analysis." He has been the principal investigator for Burleigh Corrosion Consultants, LLC, since 1993, where he conducts metallurgical failure analysis, specializing in corrosion failures.

Mr. Gary Chandler is a researcher in the Materials & Metallurgical Engineering Lab and adjunct professor at New Mexico Tech, Socorro, NM.

Mr. George Howard is founder and CEO of Restoration Systems, LLC, in Raleigh, NC. He has served as a research associate for the YDB investigation and contributing author on many of the articles published by the YDB team. Mr. Howard also publishes the "Cosmic Tusk," an influential blog concerned with abrupt ancient climate change induced by comets and asteroids.

Dr. Malcolm LeCompte is an Associate Professor and Research Director at the Center of Excellence in Remote Sensing Education and Research at Elizabeth City State University in Elizabeth City, NC. Dr.

LeCompte is one of the principal researchers and analysts for the YDB investigation and co-author of several YDB articles.

Dr. James Wittke is a Research Professor at Northern Arizona University, School of Earth Sciences and Environmental Sustainability. He has extensive analytical experience (electron microprobe, scanning electron microscope, mass spectrometry, trace element analysis) and has been involved in the YBD studies since about 2005. Drs. Wittke and Bunch have been the main team at NAU studying the textures, glass chemistry, and mineralogy of impact products (spherules, SLOs, etc.) using a scanning electron microscope (SEM) with an energy-dispersive spectrometry (EDS) x-ray analyzer to examine the material. This work requires careful calibrations to allow accurate EDS chemical analyses and carefully documenting textures and mineralogy. They have also studied nuclear-blast generated glass (trinitite) for comparative purposes

Dr. Allen West is a retired geophysical consultant for the oil-and-gas and mining industries. During his career, he worked for exploration companies in the Middle East and North and South America, searching for natural resources, including diamonds. After retirement, he helped organize a 26-member international research team of scientists who discovered evidence for a cosmic impact event 12,900 years ago that produced a discrete layer of nanodiamonds across most of the Northern Hemisphere, including Europe. During the last five years, Dr. West has published more than a dozen papers and abstracts on that impact event and the diamonds produced by the Younger Dryas Boundary (YDB) event. As a result of that research, Dr. West has developed two novel processes for creating nanodiamonds, which are the focus of an international patent application. Dr. West is one of the principal researchers and analysts for the YDB investigation and co-author of several YDB articles.

Dr. Wendy Wolbach is Professor of Chemistry at DePaul University and is one of the few experts in the world that knows how to perform the painstaking process of extracting nanodiamonds from bulk sediment. Dr. Wolbach is one of the researchers and analysts for the YDB investigation and co-author of several YDB articles.

APPENDIX E—Catastrophic Language in the OT

This investigation of the Middle Bronze Age Civilization-Ending Destruction of the Middle Ghor was inspired by my decades-long fascination with catastrophic language in the Old Testament. Many people have asked me over the years what I mean by "catastrophic language." The simplest way to define it is to give an example. Consider what the psalmist David wrote in Psalm 97:1-5 (italics emphasis mine):

¹The LORD reigns, let the earth be glad;
   let the distant shores rejoice.
²*Clouds and thick darkness* surround him;
   righteousness and justice are the foundation of his throne.
³*Fire* goes before him and consumes his foes on every side.
⁴His *lightning* lights up the world;
   the earth sees and trembles.
⁵The *mountains melt like wax* before the LORD,
before the LORD of all the earth (NIV).

When I asked my seminary professor back in the early 1980s what he thought of the phrases that I have emphasized above, he stated rather matter-of-factly, "That's just poetic hyperbole. This is a Psalm, after all!" But then I found almost the identical expression in Micah 1:3-4, the writings of a prophet that is not in poetic form (emphasis again mine):

³Look! The LORD is coming from his dwelling place; he comes down and treads the high places of the earth. ⁴The *mountains melt* beneath him and the *valleys split apart, like wax before the fire, like water rushing down a slope* (NIV).

It became clear to me that the "earth trembling" and "valleys splitting apart" was descriptive of severe earthquake activity. I thought, at first, that "mountains melting like wax" was descriptive of volcanic activity and lava flow. My first season as a volunteer excavator at Tall el-Hammam, however, caused me to change that opinion when I realized that there are no volcanic cones or fissures in the Middle Ghor from which lava might have flowed during the time of either David or Micah. That's when I realized that "mountains melting like wax" is directly tied to the "earth trembling" and "valleys splitting apart." Both David and Micah are describing landslides triggered by earthquake activity. From a distance, a landslide makes the hillside look as if it is melting! Furthermore, they are describing what they actually witnessed, using their available vocabulary within the limits of their scientific understanding of the phenomena they were observing.

King David apparently experienced great and severe geophysical and astrophysical phenomena. 2 Samuel 22 contains a song that is attributed to him in which he declares "the Earth trembled and quaked" and "the foundations of the heavens shook" (verse 8). Earth "quakes" we understand, but to say that the Earth "trembled" seems to mean something else, perhaps a wobbling or precession of the Earth on its rotational axis. If this is so, then astronomical references (the locations of stars) would also appear to shift, thus giving rise to the expression, "the foundations of the heavens shook." David also describes great bolts of lightning (verses 13 & 15) and accompanying loud thunder (verse 14) which, in the context of his song, seem to be far beyond what we would expect from the thunderstorms we experience today. He also describes (verse 16) the "valleys of the sea" being exposed and "the foundations of the Earth" (i.e., the land) being laid bare (this is exactly what happens as a tsunami approaches the shoreline) and later (verse 17) being drawn out of the deep waters.

Many different words and expressions are used throughout the Old Testament to describe and interpret catastrophic scenes. For example, the word "pestilence" (cosmic "hail"? meteorites?) is used 47 times, and the word "tumult" (shock wave?) appears 21 times. Here is a partial list of some of the ambiguous words and phrases that appear in the Old Testament (Patten, 1988):

1. Arrows, celestial	15. Hosts, Lord of
2. Besom/broom of destruction	16. Indignations
3. Brimstone	17. Lamp, burning
4. Calamity	18. Lightnings
5. Crooked serpents	19. Mildew
6. Desolations	20. Murrain
7. The Destroyer	21. Noisome pestilence
8. Devouring fire	22. Perplexity
9. Earth shaking	23. Pestilence
10. Earth trembling	24. Roaring lion, like a
11. Ensign of Destruction	25. Streams of brimstone
12. Fire falling from heaven	26. Smoke from the north
13. Flaming torches, like	27. Tempest
14. Hailstones, falling	28. Terrible shakings

29. Thunderings	32. Whirlwinds
30. Tumults	33. Woman in travail, like a
31. Vexations	

Other descriptive phrases such as "prancing," "like horses' hooves," and "melting like wax" raise the references to probable catastrophic events to over 800. Catastrophic themes appear in Genesis and Psalms and dominate certain other books of the Old Testament including Exodus, Joel, Amos, Jonah, Joshua, Judges, Nahum, Job, and especially Isaiah.

When it comes to destruction events in the Old Testament that are not the result of human agency (i.e., war), descriptions such as the infamous destruction of Sodom, Gomorrah, and the cities of the plain provide an interesting picture. The narrative description of this destruction event is contained in Genesis 19:22-28—

> [24]Then the LORD rained upon Sodom and upon Gomorrah brimstone and fire from the LORD out of heaven; [25]and he overthrew those cities, and all the plain, and all the inhabitants of the cities, and that which grew upon the ground.
>
> [26]But his wife looked back from behind him, and she became a pillar of salt.
>
> [27]And Abraham got up early in the morning to the place where he stood before the LORD: [28]and he looked toward Sodom and Gomorrah, and toward all the land of the plain, and beheld, and, lo, the smoke of the country went up as the smoke of a furnace (KJV).

The key to developing a proper understanding the biblical narrative lies in assessing the means by which the destruction occurred. Many theories exist, ranging from a sulfurous conflagration that miraculously appeared from out of the sky to various divinely-sent, but otherwise natural, phenomena such as earthquakes (accompanied by effusions of burning hydrocarbon gases), tectonic shifts resulting in a sudden lowering of the Great Rift Valley floor and covering of the cities of the plain by the waters of the Dead Sea. There is also the theory that the destruction of Sodom and Gomorrah and the cities of the plain is merely a myth and legend crafted to provide an illustration of divine judgment against sin.

Until recently, the "myth and legend" argument has dominated because no trace of the alleged cities of the plain had been confirmed. Over the past ten years, however, the Tall el-Hammam Excavation Project has provided credible evidence that the city of Sodom has finally been identified. In addition, likely candidate sites for the other named cities of the plain—Gomorrah, Admah, and Zeboiim—have also been identified. (see Collins, 2013; Collins and Scott, 2013; Collins, Kobs, and Luddeni, 2015.)

The "brimstone and fire (Hebrew *gophriyth* and *esh*) from . . . out of heaven (Hebrew *shaw-meh*)" is an apt description of the appearance of a disintegrating meteor as it creates its fiery streak through the atmosphere. The word *gophriyth* is used seven times in the Old Testament and is often translated as "burning sulfur." The word itself, however, denotes any flammable material, without chemical specificity. "Burning sulfur," "brimstone", or simply "burning stone" are all acceptable translations of *gophriyth*. All of these terms would be equivalent in the minds of MBA people when trying to describe a meteor.

That the *gophriyth* and *esh* is described as coming from "out of heaven" need not be construed as having come from the place where God dwells (the literal "heaven"). The word *shaw-meh* is usually used to describe both the visible arch where the clouds dwell (i.e., the atmosphere) and higher and greater visible arch where the celestial bodies move. The phrase "out of heaven" should therefore be understood as meaning "from beyond Earth's atmosphere" (i.e., from outer space), the place from which all meteors originate.

That the writer of Genesis (traditionally thought of and also affirmed in the New Testament as being Moses) would identify "the LORD (Hebrew *YHWH*)" as the source of the *gophriyth* and *esh* is consistent with the unique Hebrew perspective that God is the creator of all things who uses his creation to accomplish his purposes. (By contrast, other contemporary nations observed the cosmos and made gods out of what they saw.) It is a simple matter for the Maker of the great constellations (Job 9:9) to alter the course of an asteroid and transform it into a meteor on a collision course with the Earth. He needs no magic to create destruction out of thin air, even thought that may be what it looked like to the eye witnesses!

The phrase "he overthrew (Hebrew *haphak*) those cities" is interpreted by some to mean that God literally flipped the landscape over like a giant pancake, thus burying the cities of the plain beneath the surface of the ground. Although this sounds like a plausible interpretation, it fails to account for the presence of numerous visible and accessible occupation mounds, including Tall el-Hammam, that bear witness to a once thriving and robust Middle Bronze Age civilization. If the landscape had been literally flipped upside-down, then these sites would not be where they are today! Of the 92 appearances of the word *haphak* in the Old Testament, it is translated "overturn" only when referring to a city, including Sodom and Gomorrah and the cities of the plain. In the majority of its appearances, *haphak* is translated "change" (as in appearance or state) or "turn" (as in change color). All of these concepts apply to the destruction of Sodom and Gomorrah and the cities of the plain. These cities were utterly destroyed, reduced to dust and ruins, and a major portion of exposed buildings and their contents were either blown into the surrounding hills or sucked up into the stratosphere and beyond when the meteor traveling at hypersonic speed punched a hole in the atmosphere and momentarily brought vacuous outer space down to the surface of the Earth.

The destruction was not limited to a few select urban centers, but involved "all the plain (Hebrew *kikkar*)." Although appearing in variant forms in all of the Semitic languages, *kikkar* is used only in the Hebrew Old Testament (and only 13 times out of 69 appearances in 55 verses) to refer to a piece or real estate, namely, the circular widening of the Jordan River Valley at the north end of the Dead Sea. The entire *kikkar* was affected by this destruction event—the urban centers, the "inhabitants of the cities," and the landscape itself. As I documented in my main discourse, the analysis of soil from Tall el-Hammam indicated that the topsoil and "that which grew upon the ground" were stripped away, and the remaining sub-soil was contaminated by a super-heated brine spray of water from the Dead Sea.

Contrary to the opinions of most biblical scholars, I do not believe that Lot's wife was turned into a solid "pillar of salt" when she looked back one last time toward Sodom. I believe instead that she was (unfortunately for her) in an exposed place when she looked back and was overtaken by a plume of super-heated brine spray. As a result, she literally died on her feet and was coated with a salt encrustation that gave her the *appearance* of a pillar of salt. As her body collapsed, so too did her salt cocoon, leaving nothing behind but an interesting anecdote to the destruction event.

The destruction event must have occurred late in the day (see Appendix F for a full discussion of the timing of the destruction event), because "Abraham got up early the next morning . . . and looked toward Sodom and Gomorrah . . . [and saw a great column of smoke that looked like] the smoke of a furnace." The smoke must have been both thick and black as the remaining organic building materials and contents of the buildings continued to smolder and burn. Although limited to just five short verses, this was truly a great narrative description of an amazing destruction event!

The primary purpose of my dissertation on "The Middle Bronze Age Civilization-Ending Destruction of the Middle Ghor" was to document the existence of a thriving civilization in the Middle Ghor during the Middle Bronze Age and offer possible explanations for the mechanism of its demise based on physical evidence, why the area remained unoccupied by permanent settlements for over 600 years, and what changed to allow civilization to return. This exercise cannot explain every occurrence of catastrophic language in the Old Testament, but can at least help to explain one of the more infamous catastrophic events in the Old Testament. If Tall el-Hammam is indeed Sodom, then we now have a reasonable explanation for the "brimstone and fire" that rained down upon Sodom and Gomorrah and the cities of the plain.

References:

Collins, S.

2014   *The Search for Sodom and Gomorrah.* Albuquerque: Trinity Southwest University Press.

Collins, S. and L. C. Scott

2013   *Discovering the City of Sodom.* New York: Howard Books, Division of Simon & Schuster.

Collins, S., C. M. Kobs, and M. C. Luddeni

2015   *The Tall al-Hammam Excavations: Volume One—An Introduction to Tall al-Hammam with Seven Seasons (2005-2011) of Ceramics and Eight Seasons (2005-2012) of Artifacts.* Winona Lake, IN: Eisenbrauns.

Patten, D. W.

1988   *Catastrophism and the Old Testament.* Seattle: Pacific Meridian Press.

APPENDIX F—The Timing of Sodom's Destruction

Note: This appendix contains an article entitled "A Preliminary Estimate of the Timing of Sodom's Destruction" that I wrote and was published by the Society for Interdisciplinary Studies in the journal Chronology & Catastrophism Review, 2015:1, 34-37. I have updated this presentation with my current thinking on the subject.

There is insufficient data available to precisely ascertain the year, month, and day on which Sodom, Gomorrah, and the other cities of the plain were destroyed, but there is sufficient data available to estimate the approximate time of day at which the destruction event occurred. The purpose of this article is to examine these literary, linguistic and archaeological clues.

The only primary source document that we have on the destruction of Sodom and Gomorrah and the Cities of the Plain is the Bible. The destruction event is described in the book of Genesis, Chapter 19. Many words have been written in numerous attempts to interpret the text and identify the location of Sodom and the mechanisms of its destruction, but few words have been offered regarding the time of day at which the destruction event occurred. Nevertheless, there are clues in the text which strongly suggest the time of day, and certain discoveries during the excavation of Tall el-Hammam appear to confirm these textual clues.

The narrative of the text is contained in verses 1 through 28 and spans a period of roughly 36 hours from the evening of "Day 1" (verse 1) to the morning of "Day 3" (verse 27). The actual destruction event occurs on "Day 2" (verse 15) at about the time when Lot reached Zoar (verse 23). The question is: What "time" is that? The Hebrew text states: *hashemesh yatsa' al-ha'erets*, which is variously translated "the sun was risen upon the earth" (KJV[1]), "the sun had risen on the earth" (RSV[2]; ESV[3]), "the sun had risen over the land" (NIV[4]), etc. This phrase has been broadly interpreted to mean anything from "the full orb of the sun had cleared the horizon" (i.e., early morning) to "the sun was at its zenith" (i.e., high noon).

---

[1] King James Version, 1611.
[2] Revised Standard Version, 1902.
[3] English Standard Version, 2002.
[4] New International Version, 1978.

Many proposals have been made concerning the location of Zoar. I personally prefer the general location shown on the famous Madaba Mosaic Map, which is in the vicinity of the Arnon River (Wadi Mujib) on the east shore of the Dead Sea near the southern end of the deep north basin.[5] The cliffs that rise from the east shore of the Dead Sea in this location are very tall and almost vertical. It is nearly noon before the full orb of the sun clears the cliff and directly illuminates the eastern shoreline! But, is that what the writer wants us to understand—that the destruction of Sodom occurred at noontime? I think not, and I shall endeavor to make my case through a detailed examination of the text and correlation with evidence which I personally found while excavating at Tall el-Hammam.

What does the text say? The Hebrew word *shemesh* appears in six verses in Genesis and always means "sun".[6] There are no interpretation issues with this word; its translation is consistent. The word *erets* appears in 252 verses in Genesis. Depending upon the context in which it appears, *erets* is translated "earth" (as a general reference to the planet as a whole) or "land" (as distinct from the water, or the habitation of air-breathing creatures including man). Thus, there are no interpretation issues with this word, either. The confusion—which comes from a lack of specificity in the English translations—arises with the word *yatsa'*, which appears in 75 verses of the 51 chapters in Genesis. Eight of these verses are in chapter 19.

Ignoring chapter 19 for the moment, how is *yatsa'* used and translated in the rest of Genesis? In all but one of the 67 verses in which *yatsa'* appears in Genesis, other than in chapter 19, the word is used in the context of (a) a person or persons coming forth from or out of somewhere in order to go somewhere else (significant majority of appearances); or (b) bringing something forth from or out of something or somewhere (minority of appearances). Thus, the general sense of *yatsa'* is to go out from somewhere to another location, which typically involves passing completely through a building's doorway or a city's gate. The one "odd"

---

[5] Note that the Madaba Map was created during the Byzantine period when the Dead Sea was near its record low water mark, as it is today. The map shows neither the southern basin of the Dead Sea (which was dry) nor the Lisan (the peninsula that separated the northern and southern basins). Failure to recognize this feature of the Madaba Map has led many to incorrectly identify Zoar with Bab edh-Dhra (Neev and Emery, 1995) or Safi (Glueck, 1926).

[6] *Shemesh* is also the name of the Canaanite sun god.

204

translation is in 42:28 where *yatsa'* is used in connection with the "fail-ure" of the heart/emotions in a moment of extreme stress. Even here, the phrase could be translated "their hearts *went/gave out*" instead of "their hearts *failed* (KJV)/*sank* (NIV)"; thus, the general meaning of the word *yatsa'* is preserved.

In the Sodom narrative of Genesis 19, *yatsa'* is translated with the same general meaning of moving people through a building doorway or the city gate in verses 5, 6, 8, 12, 14, 16, and 17. In verse 23, a unique and totally different phrase is used to translate *yatsa'*: "By the time Lot reached Zoar, the sun **had risen over** the land."

The narrative of chapter 19 begins in vs. 1 with the evening (Day One) appearance of two angels[7] at the city gate of Sodom where they found Lot sitting, as was the custom of the city elders. Presumably, Lot would have been sitting on the "bench" in the city gate (which was found in 2012 by the team led by TeHEP Assistant Director, C. Kobs). Lot then convinced the two visitors to spend the night with him in his house.

Me, sitting on the "bench" in the main gate of Tall el-Hammam, 2012

Their evening is interrupted when the men of the city demand that Lot <u>bring out</u> (*yatsa'*, vs. 5) the two visitors. Lot <u>went outside</u> (*yatsa'*, vs. 6) and offered to <u>bring out</u> (*yatsa'*, vs. 8) his two young daughters instead. After Lot's visitors struck all of the men outside with blindness (vs. 11), they instruct him to gather up his kin and <u>get</u> them <u>out</u> (*yatsa'*,

---

7   The Hebrew word is *malak* which generally means "messenger." Because these "messengers" later strike blind the people trying to break down Lot's door (see 19:11), they are thought to be messengers from God, hence, "angels."

vs. 12) of the city. So Lot <u>went out</u> (*yatsa'*, vs. 14) and spoke to his [soon-to-be] sons-in-law, but was unsuccessful in convincing them to leave with him. Thus ended Day One.

At dawn the next day (Day Two, vs. 15), the two visitors try to hurry Lot, his wife and two daughters out of the city. They dragged their feet, however, so the visitors <u>brought</u> (dragged?) them <u>out</u> (*yatsa'*, vs. 17)—out of the house or out of the city is not clear—with instructions to flee to the mountains. Presumably, the visitors intended that Lot and his family flee to Mount Nebo, visible from Tall el-Hammam/Sodom, and accessible over a long and rugged vertical climb of almost 3,000 feet over a distance (by trails, not "as the crow flies") of about 15 miles. Apparently fearful that the destruction of the city could come at any moment, Lot pleads with them to let him and his family flee to Zoar, a small and insignificant (which is what "Zoar" means) town instead (vss. 18-22). This journey would be downhill by about 800 feet from Tall el-Hammam/Sodom to the northeastern corner of the Dead Sea, and then relatively level along the narrow shoreline on the eastern side of the sea. The distance from Tall el-Hammam/Sodom to Zoar is about 30 miles and can be walked in about 10-12 hours.

All seven appearances of *yatsa'* in chapter 19 thus far have been translated into English with the common meaning of going into or out of a building (i.e., passing completely through a door-way) or the city Sodom (i.e., passing completely through the city gate). In vs. 23, however, virtually every English translation uses the same phrase to translate *yatsa'*—"when the sun had <u>risen over</u> the earth". The question remains: What does this mean?

I have already mentioned the two most common understandings of this phrase—that the sun had cleared the horizon, or that the sun had reached its zenith. If we use an idiom that is more closely aligned with the common sense of *yatsa'*, this phrase should read "when the sun had <u>passed over</u> the earth." *Passed over* [the earth] is equivalent to *gone out* or *passed through* [a doorway or gate]. In other words, the sun was *setting*, not just risen or at its zenith in the sky. Thus, Lot and his family arrive at Zoar as the sun is setting (vs. 23), which is consistent with a 10-12 hour walk from Sodom to Zoar.

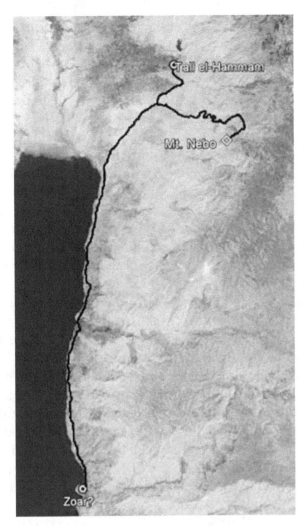

Lot's escape route alternatives

Continuing the analysis of the text . . . At that same time, destruction rained down upon Sodom and Gomorrah and the other cities of the plain (vss. 24-25). Lot and his daughters would have been shielded by the cliffs surrounding Zoar from the destructive force. Lot's wife had lagged behind seeking one last glimpse of Sodom, however, and consequently died (vs. 26). (Ironically, the cliffs along the eastern shore of the Dead Sea blocked her view as soon as she turned south at the northeast corner of the sea.) Thus ended Day Two.

The next morning (vs. 27; Day Three), Abraham looked down toward Sodom from the vicinity of his encampment on the plains of Mamre (see 18:1). Mamre is near Hebron, about 20 miles south of Jerusalem and 15 miles west of the Dead Sea. Although Abraham's camp was about 3,000 feet above sea level, his direct view of Sodom and the other cities of the plain (500-1,200 feet below sea level) was blocked by the higher spine of the Judean hills. Therefore, to look *down* upon Sodom, he must have climbed to the crest of hills near his camp. What he saw is described as "smoke from a furnace"—i.e., a large column of thick, black smoke rising from the plain (vs. 28). Thus ends the narrative.

Does a *setting* sun make sense with the timeline of the narrative? Yes! First, Lot's exodus from Sodom begins at dawn with preparations to leave the city with his immediate family. He makes haste slowly, however, and his two visitors are compelled to forcibly escort them out of the city. The first light of dawn would appear in the Great Rift Valley where this drama took place between 5 and 6 AM. If the packing and negotiations regarding their destination were accomplished within an hour, it was possibly as late as 7 AM before Lot and his family actually left Sodom. Assuming that they walked at a brisk pace to Zoar, considering the circumstances of their departure, it would have taken them about 10 hours to get to Zoar, thus arriving around 5 PM, which is approximately the time (with some variation, depending upon the time of year) that the sun would drop behind the spine of the Judean Highlands as viewed from Zoar. Second, the low, late-day sun would still be shining at Abraham's location when the destruction hit Sodom and the other cities of the plain. There is no information in the text to indicate what, if anything, he might have seen or heard that evening. There may have been a glow in the sky to his northeast as the consuming fires illuminated the billowing smoke from underneath. At any rate, it was too late in the day to go out for a look, so he waited until first light the next morning.

From this detailed analysis of the language and story line of the text, it becomes clear that the destruction of Sodom and Gomorrah and the other cities of the plain occurred late in the day, probably at or immediately after the sun set below the western rim of the Great Rift Valley. But, can this be confirmed from the archaeological data?

Again the answer is: Yes! During Season Seven (2012) of the Tall el-Hammam Excavation Project, I was assigned the task of excavating a trench inside the newly discovered tower of the main city gate complex. The purpose of the task was to determine the number of remaining courses of mudbrick as well as the number of courses of underlying foundation stones. I excavated along the interior face of the west wall into the southwest corner of the tower. As I approached the corner, I encountered the sherds (broken pieces of pottery) of a broken cooking pot intermingled with bones from a small kid goat. Finding broken cooking pots at the site is not unusual—we find them all over the place, and especially in a communal cooking area not far from the tower. Two things struck me as unusual about this pot. First, its placement in the corner of the tower was apparently purposeful since there was no evidence of cooking near it or immediately beneath it. Second, the pot apparently contained an un-touched meal made with goat meat immediately prior to its destruction.

Excavation of the Gate Tower

A meal containing goat meat is typical of an evening meal, not a morning meal. The placement and appearance of the broken cook pot led me to the conclusion that this was dinner for the tower guards. Someone had carried it into the tower from the nearby communal cooking area and set it in the corner for them. Before the guards had the opportunity to come down to eat their evening meal, sudden destruction fell upon them, their tower, and the entire city. The pot was crushed by falling debris, and only the potsherds and goat bones remained for me to find.

Whether or not goat bones were found intermingled with sherds of cooking pots elsewhere at Tall el-Hammam during previous seasons I cannot say. After pointing out this find to Director Dr. Collins, however, other broken cooking pots mingled with goat bones were noted during Season Eight (2013).

It is apparent from the discovery of an untouched evening meal in the destruction debris of the gate tower that the destruction event occurred at about the time of the evening meal—near sundown.

Conclusion: While there is insufficient data available to precisely ascertain the year, month, and day on which Sodom, Gomorrah, and the other cities of the plain were destroyed, there is sufficient data available to estimate the approximate time of day at which the destruction event occurred. An analysis of the language and story line of Genesis 19:1-28, the only available written text describing this event, reveals that the event occurred late in the day, probably around sunset. Archaeological data from Tall el-Hammam/Sodom also point to a late-in-the-day destruction event.

Although the nature of the destruction is not germane to the subject of this article, the text that was analyzed herein states that the destruction came from the sky. Archaeological evidence from both Tall el-Hammam and neighboring sites suggest that this destructive force approached Tall el-Hammam from the southwest. If this is indeed the case, then the destructive force came out of the direction of the setting sun. Thus, the inhabitants of Sodom and Gomorrah and the other cities of the plain *never saw it coming*!

References:

Glueck, N.
  1935 "Explorations in Eastern Palestine, II." *AASOR* XV for 1934-1935.

Neev, D. and K. O. Emery
  1995 *The Destruction of Sodom, Gomorrah, and Jericho: Geological, Climatological, and Archaeological Background.* Oxford: Oxford University Press.

CPSIA information can be obtained
at www.ICGtesting.com
Printed in the USA
BVHW040453290921
617681BV00018B/1239

9 780692 609699